Common Riparian Plants of California

Common Riparian Plants of California

A FIELD GUIDE FOR THE LAYMAN

Phyllis M. Faber and Robert F. Holland

Drawings by Virginia I. Dains

PICKLEWEED PRESS

PICKLEWEED PRESS
212 Del Casa
Mill Valley, California 94941

First Edition
First Printing 1988

Printed in the United States of America

ISBN 0-9607890-1-4
Library of Congress Catalogue Card Number: 88-092430

Other publications from Pickleweed Press:

Common Wetland Plants of Coastal California 1982

ACKNOWLEDGEMENTS

Several people have added to this fieldguide through their knowledge or by their various skills; we are most grateful to each of the following very special group: Bob Holland's wife, Ginny Dains has drawn the charming scale drawings and the riparian cross-section, and has provided good cheer, review, and middle-of-the-night food; Larry Heckard of the Jepson Herbarium has checked plant identifications and made helpful suggestions; various people have contributed to the plant collection in varying ways including Pat Flannagan, Crawford Cooley, Terry Sholars, Steve Edwards, Tom Griggs and Bruce Handley from The Nature Conservancy, Nancy Wymer, Ginny Dains, and Bob Holland; Wilma Follette has reviewed the plant descriptions and has made many useful additions; Larry Stromberg, Barry Hecht, and Susan Sanders have reviewed the text and have made numerous helpful suggestions; Sally Porter has provided ever-helpful typing assistance; Joan Browning is responsible for book design and welcome counsel, Phoenix Word & Press for typesetting, and Braun and Brumfield for printing. We thank them, one and all.

DEDICATION

This book is dedicated to those who appreciate the remnants of the once great riparian community of California and who devote their energies towards its preservation and restoration for present and future generations.

TABLE OF CONTENTS

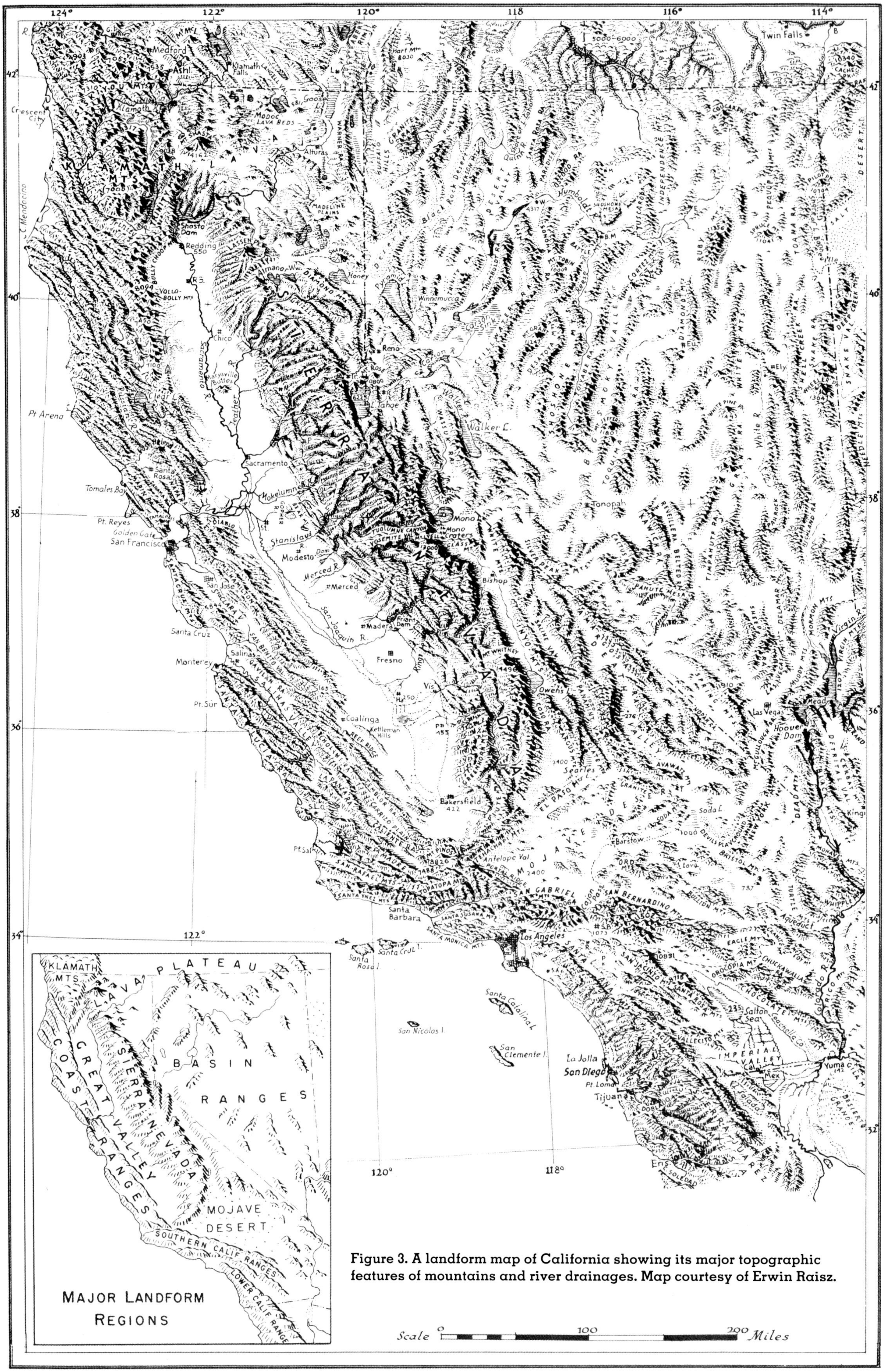

Figure 3. A landform map of California showing its major topographic features of mountains and river drainages. Map courtesy of Erwin Raisz.

ABOUT

Riparian Habitat

Did you ever wonder why is it that the lakes and rivers are fresh instead of salty? Or why rivers don't just run dry? And what do these questions have to do with the specialized kinds of plants, riparian vegetation, found beside streams? The answers lie in a physical-chemical process known as the hydrologic cycle.

What is the Hydrologic Cycle?

The earth is a watery planet. About 80 percent of its surface is covered by ocean. It is impossible to imagine life on earth without water; it is an essential part of every living cell. But ninety-seven percent of the earth's water is in the oceans; only three percent is fresh. Most of the world's store of fresh water sits in ice packs, snowfields, lakes, reservoirs, and groundwater basins, with only a fraction flowing in streams. Yet flowing water, obviously, is one of the dominant forces that shapes the land.

Streams flow into lakes or oceans from which water evaporates, making it once again available to condense into precipitation. This solar-powered process of repeated evaporation, condensation, and precipitation is termed the hydrologic cycle. It is an on-going process worldwide that leads to progressively saltier oceans, but guarantees that rain water will always be fresh.

Streamside habitats are so strongly influenced by stream flow that they invariably are different from the habitats on surrounding uplands. Stream flow of course varies with the vagaries of precipitation. Floods, droughts, and seasonal variations in stream flow or groundwater levels all affect where riparian habitats may occur and the types and densities of species that flourish within them.

What Makes Streams Flow?

Water drains slowly out of a saturated sponge. The soil acts like a sort of sponge, holding enormous quantities of water for days or weeks after a storm. Plants draw upon this water. But gravity also is drawing the water, only downward. Much of the water is gradually released from the soil into the stream system. This deferred release of soil water, together with ground water returning to the drainage net through springs and seeps, is what keeps our streams flowing even when it hasn't rained for months.

We take for granted that some streams are "larger" than others. Our literature is full of references to mighty rivers and babbling brooks, fanciful images laden with size comparison. These variations in size arise through the interactions among several important variables including climate, watershed size, the "leakiness" of the underlying geology, the soil mantle, and the vegetation in the watershed.

Climate obviously controls the amount of water a watershed will produce. Mean annual rainfall in California ranges from 2-3 inches at Furnace Creek in Death Valley to over 110 inches per year at the town of Monumental in the rainforests of the Pacific Northwest. These differences are reflected in the net flows of local streams. Death Valley has dry desert washes subject to intense but infrequent cloudbursts. Even the smallest valleys in the Northwest have perennial streams maintained by precipitation.

The timing of the rainy season is nearly as important as the amount of rainfall; most of California receives the bulk of the year's precipitation between October and April, with relatively dry, warm summers. Plant moisture stresses are lower in the winter because of lower temperatures, but growth potentials also are reduced because of the lower temperatures and cloudy skies. These factors combine to reduce the portion of precipitation lost to plant evapotranspiration, leaving more water to drain out of the soil mantle over the winter.

The effect of watershed size on stream flow is direct. Larger areas can gather more moisture than smaller ones. There simply is no comparison between the flows of the Colorado River and, say, the San Gabriel River. The former is fed by several western states, the latter by a part of Los Angeles County. Riparian habitats in smaller watersheds are subject to abrupt changes. Some of these changes are natural episodic or periodic events, such as floods or massive sedimentation and channel changes following large wildfires or landslides. Riparian areas in smaller watersheds also tend to be disproportionately affected by the hydrologic changes and increased sedimentation associated with timber harvesting, overgrazing, or other human activities.

Every watershed has an upland soil cover that has developed in part through weathering of the underlying basement rock. In California, this bedrock varies from massive, solid granite such as in Yosemite to the fractured, erodible melange of the Franciscan formation of the Coast Ranges. These kinds of differences in bedrock geology can have profound influences on stream flows. Massive bedrock presents few opportunities for water to escape from the soil column and drain away to depth. Bedrock that has been fractured, stretched, or faulted presents many more avenues for ground water movement. Each of these possible routes can affect the proportion of precipitation available for stream flow and when it may reach the stream.

The Riparian Experience: Deposition and Erosion

Streams move more than just water. Sediments, "salts" (dissolved solids), woody debris, and other organic material also are carried downstream. Larger sediments and woody debris tend to be transported primarily during high flows. At such times streams may erode their banks or they may pour over their banks into the adjacent floodplain. These overbank floodwaters leave fresh deposits of sediment (alluvium) everywhere within the channel and overbank areas near the stream. But they also can remove the same material. There is a tradeoff between erosion and deposition, the balance between them varying with distance from the riverbank.

Smaller floods may remove some, but not all of the alluvium deposited during larger floods. Floodwaters that overtop the bank flow into the surrounding riparian woodland, depositing sand and silt transported by the stream. Additional material accumulates with each flood, gradually increasing the site's elevation. As much as a foot or more of material may accumulate in a single flood. As vertical accretion occurs, the site will be less frequently or less intensely flooded. On the other hand, a channel also may erode away entire bank segments, lowering the level of the floodplain in irregular patches. Likewise, a stream may move laterally during a flood, potentially changing its course and leaving a site abandoned at some distance from the water's edge. At some distance from the bank, material accumulates most rapidly, developing into low, broad, natural levees that will be overtopped only during extreme events. Though difficult to detect in today's man-altered environment, these low natural levees are in some cases many miles wide.

Figure 1. Vertical air photograph of a segment of the Sacramento River in Butte and Glenn counties. The present river course appears as a dark, sinuous band through the center of the image. Several previous channels also are visible. The line separating the two counties indicates the river's position at the time the counties were created. Numbered dots indicate miles above the mouth of the Sacramento River at Collinsville and can be used to show the extent the river has changed course since this survey was conducted. Thus, several changes in the river's course are shown to have occurred during recent historic times. Several oxbows such as seen near Eddy Lake indicate other, older alignments of the river. Image, taken 22 May, 1984, is reproduced here through the courtesy of the US Army, Corps of Engineers, Sacramento District.

What is life like for plants living beside a stream?

Several riparian plant species such as willow, cottonwood, seepwillow, or alder require clean, freshly deposited, moist bare sand for germination. These species all time their seed release to coincide with the most likely time such conditions would be encountered, just after the normal time of high water.

Pioneer colonizers compete intensely for opportunities to grow in the same limited space — sunny places with freshly deposited soil. These species each have very rapid growth rates; many are capable of reaching three or five feet in a single growing season as they race to occupy the available space. The hot summer sun puts plants under intense heat loads. Plants that cannot reach life-sustaining water supplies deep within the alluvium soon wilt and die. Plants that do establish successfully still face the prospect of pruning, excavation, or burial during the next flood event. As a consequence, most such species have strongly developed vegetative reproduction. Once established, their roots can withstand remarkable trauma and resprout vigorously once the waters have receded. Even detached pieces of the stems or roots of many of these species will quickly develop new roots or shoots if they are buried in suitable locations. Sandbar willow derives its name from its preferred habitat and can prosper under all of these types of disturbances.

Once plants become established, they begin to influence their environment. The presence of vegetation increases turbulence and may slow the velocity of floodwaters, thereby reducing the stream's ability to transport suspended materials.

The current deposits sediment in the canopies of and downstream from established shrubs. Both erosion and deposition occur during every flood, of course. These processes foster episodic plant reproduction, tending to result in whole cohorts of like-sized individuals, established following some particular disturbance event.

Initial colonizers require sunny, bare mineral soil for germination; however, their presence on a site virtually assures that they will not be able to replace themselves in that same site. Instead, they provide appropriate germination conditions for a more varied assortment of plants that require shade during establishment. This group includes several trees such as box elder, ash, and walnut as well as vines such as wild grape, blackberry, and poison oak. The site still will experience flooding virtually every year, but the floods will have less effect on the denser, deeply rooted, and layered vegetation. Barring total removal by a realignment of the river, this site becomes a part of the floodplain and will continue to support a broad array of trees and shrubs.

Floodplain forests are as diverse and productive as any ecosystem in California. The mixture of shade-tolerant species reduces but does not eliminate a kind of patchiness observed closer to the stream. Sun-loving species may persist for a few decades following establishment, all the while being joined by new neighbors. This more species-rich and structurally diverse canopy offers a broader spectrum of resources to wildlife, so it supports many more animal species.

The understory in these stands also is more species diverse, often to the point of being an impenetrable tangle of plants that shout with a tropical exuberance—a riot of leaves everywhere festooned with long draping vines. While this environment provides amenities to many wildlife species, those of us who try to observe them often are frustrated by the density of the canopy—that yellow warbler may be only ten feet away, but he is entirely hidden by the foliage. The density of the tangle is in large part a reflection of the reduced intensity of pruning experienced during high flood flows.

Flood intensities generally are lower in this outer floodplain. There is less physical disruption from erosion and battering by floating debris. The principal outcome is accumulation of fine, fertile alluvium. These sites are relatively free of rejuvenating disturbances, so those species requiring such disturbances gradually are eliminated from the community. Longer-lived trees with shade-tolerant seedlings eventually come to dominate a site. This was the setting of the vast, productive forests of valley oak that once lined the major water courses of the Great Valley.

What do riparian plants have in common?

Despite of the profound range of climatic conditions found in California, several common themes distinguish riparian vegetation from the typical upland vegetation found within any climatic zone.

There are marked physical and floristic similarities among riparian communities when streamside habitats from disparate climates are compared. Foremost among these is the ready availability of water during the sunny summer season. This difference is obvious in desert areas, where one may find oases of willows, cottonwoods, and fan palms along spring-fed waterways standing in stark contrast to the surrounding miles of desert scrub; however, the distinction is no less important in the temperate rain forests of northwest California, where, historically, the largest trees are logged from streamside floodplains. Dense forests once occurred in the outer floodplains of the lower Colorado River, but there the climax community was a bosque dominated by screwbean mesquite and a variety of shrubs.

The convergence of life forms seen in riparian habitats also reflects adaptations to flooding disturbance. A stream may flood every year, or it may flood every few decades; the frequency of flood disturbance shapes the vegetation.

Figure 2. Idealized cross section of an alluvial stream. The asymetrical cross section of this river channel suggests that the river is making a gradual turn to the left. The channel is deeper and carries faster moving water on the outside of the bend. Here the river is cutting into the base of an older alluvial terrace. Valley oaks grow perched above the river on the right side. At the river's right bank, a narrow band of alders take advantage of small cracks and tumbled blocks of soil to root at the water's edge. The coarse cobbley material on the river bottom, or "bedload", moves downstream during floods. The current in the shallower side of the channel flows slowly and fine particles, sand and silt, are deposited. Emergent herbs grow in the quiet water of a "back eddy". Over time, the river has cut many feet into the cliff and deposited a series of progressively older and higher "scroll bars" that now support lush riparian vegetation. Nearest to the water's edge, a narrow grassy beach gives way to a dense thicket of sandbar willow. Frequent inundation during high flows subjects plants in this site to intense pruning as well as sporatic deposition. Vegetative reproduction is favored. Few trees survive the regular and intense disturbance found at the water's edge. Further from the river, flooding is less severe, and trees can become established. Cottonwoods or tree willows are the first pioneer species to form closed canopy forests. Eventually, shorter-lived pioneers are replaced by a suite of riparian trees, such as boxelder, ash, walnut, oak, and a profusion of vines dominate the canopy. Abandoned or high water channels may be mantled by bramble, buttonwillow, or young trees that struggle to establish in the narrow gap of light.

The fresh alluvium left behind by a flood sets the stage for those colonists that start the successional sequence. Desert washes and redwood-shrouded north coast streams share this attribute; establishment is mediated by flooding events. Usually this leads to patchy stands of more or less even-aged individuals established after each flood event.

Many of the most important riparian trees and shrubs are most closely related to species that grow in the deciduous forests of eastern North America. Most of California's maples, ashes, walnuts and dogwoods are found in moist settings. These species share the deciduous habit with cottonwoods, sycamore, and all the willows that line streams. These plants require abundant water supplies throughout their growing season. In the eastern deciduous forests, this demand is met by abundant summer rains. In California, it is met instead by streamflow. Uplands may be mantled by grassland, chaparral, oak woodland, or conifer forest, but the streams usually have distinct canopies dominated by winter-deciduous species with broad and soft-textured leaves. In most parts of the state, these vegetative features create a clear visual distinction between the riparian corridor and adjacent upland habitats.

Why is riparian habitat so vital to animals?

Flowing water's influence on streamside communities extends to the animal community as well. Riparian habitat is extremely important to wildlife. Some animals, such as beaver, aquatic garter snake, yellow-billed cuckoo, or willow flycatcher, depend on the resources of the riparian zone for their entire life cycle.

A much longer list of animals require streamside communities for one or several aspects of their habitat needs. For example, Swainson's hawks forage in grasslands, but they require large oaks or cottonwoods near streams for nesting. Over half of the breeding birds in many areas are associated with riparian habitats. Forty percent of the mammal and reptile species in the western United States depend on riparian habitats for some or all of their needs. For amphibians, the figure is over eighty percent.

Migrating birds use riparian habitats as corridors, providing shelter, food, and water as the birds work their way north in spring and south in the fall or, back and forth between different habitats daily, or day and night. Migratory mammals, such as deer, also rely on these habitats for passage between the valley floor in winter and their summer mountain haunts.

The terrestrial plants of the riparian community are critical to the function of the assemblage of animals in a stream ecosystem. They are the base of a food web found in rivers and streams. Fish use overhanging vegetation for cover, and use flooded vegetation for spawning. Shade from cottonwoods, alders, and willows keeps stream temperatures cool in the summer, which is extremely important in the development and growth of young fish. Leaves and other debris that fall into the stream provide an important source of nutrients for insects and other stream invertebrates. Most aquatic insects, which are essential prey items for fish or other predators, spend part of their life cycle in terrestrial riparian vegetation. Streamside vegetation intercepts sediments eroded from surrounding uplands, sediments that otherwise would cloud the water and clog fish spawning gravels.

Riparian settings attract wildlife for the same reasons they appeal to humans; they offer a favorable microclimate combined with abundant resources. Rich alluvial soils and plentiful water supplies during the growing season result in very high plant growth rates. The variety of trees, shrubs, and herbs results in a complex forest canopy that offers many alternative perching and nesting sites, as well as places to hide from predators or unfavorable weather. The variety of plants growing along streams provides a long menu of food items; foliage, fruits, seeds, and wood all form the basis for some animal's diet. Insects abound and the food web can be large and complex by the time the top predators visit a riparian site.

Not only is productivity high, but the resources are distributed in a patchy fashion that reflects the river's habit of changing direction midstream. Flood-induced stream realignments usually create linear, homogeneous stands of vegetation with sharp boundaries between adjacent patches of different species composition. These ecotones, or zones of transition from one plant community to another, combine the resources of several communities in a small area. For example, a northern oriole may nest in a forest tree and forage for insects in an adjoining willow thicket. The same bird may gather nesting materials from a nearby emergent wetland. Recurrent flooding maintains this habitat complexity, which is so important, over the long term.

ABOUT

The Status of Riparian Habitat

How Much Riparian Habitat is There in California?

John Fremont's accounts of his explorations of the early nineteenth century Great Valley are replete with references to "fine woodlands" of valley oaks that spread over "wavering fields" of grasses, especially between Sacramento and Stockton. By the time he reached the Kaweah River his opinion of Great Valley riparian habitats must have changed, because he was thwarted in trying to get through the riparian jungle there. Instead, he was forced to turn west into the "tulares", the emergent wetlands surrounding Tulare Lake.

Streamside settings have a universal attraction to people, especially in the west. Riparian features figure prominently in California place names. Lots of streams have names like Sycamore or Ash Creek. Others have the same names, only in Spanish as Aliso or Fresno Creek. Even the Spanish and Mexican land grants refer to riparian situations, for example, the Sanjon de Santa Rita grant along the San Joaquin River. Sanjon is a prevarication of "zanja", a deep, sluggish ditch or canal; it seems to refer to a slough or quiet waterway such as the San Joaquin must have been as it debauched into the grasslands of central Merced County. Three other land grants in Monterey, Fresno, and Sacramento counties have similar names.

Water available in an otherwise dry climate made such sites among the first to be settled. Homesteads became villages became cities as first agriculture then population spread across the land. As a result, most western states have experienced dramatic reductions in riparian habitat; estimates range around 80-90 percent.

Estimates of just how much riparian habitat might have existed in California are somewhat elusive. Several studies have tried to estimate the original extent of riparian habitat in various parts of the state. Each has used different assumptions and has studied a different area of California, although most studies have focused on one or another part of the Great Valley.

The earliest objective estimates of the extent of riparian corridors have been gleaned from early surveyors' records. According to their reports, the riparian forests along the lower American River near Sacramento stretched as much as four miles back from either side of the stream. Similar gallery forests bordered most Great Valley and Coast Range streams.

Riparian zones were mapped by the earliest soil surveys as distinct soil types, usually accompanied by enticing descriptions of dense thickets and forest that regularly experienced significant floods. For example, L. C. Holmes described the soils now designated as Hanford soils as "originally covered with a heavy tree growth, willow, cottonwood, sycamore, and many other trees of this class occupying the river bottoms, while oak predominated on the delta land" in a 1913 reconnaissance of the soils of the Sacramento Valley.

Currently available data suggest that riparian habitats once covered some 920,000 acres in the Great Valley alone. This is more than fourteen hundred square miles, or about the size of San Joaquin or Kings County. Agricultural and urban development, flood control, and bank protection projects have reduced this drastically. Only about 12,000 acres of relatively pristine habitat still existed by the end of the 1970s, a loss of nearly 99 percent.

No one has estimated how much riparian habitat existed in the Los Angeles basin at the time of Spanish contact, but virtually none remains today. Severe flooding during the 1930s prompted construction of a massive system of dams and stream "improvements" that either removed riparian habitats outright or allowed development in what formerly had been floodway. Most streams in the basin were realigned and converted to concrete-lined culverts, void of vegetation. The few plants that do manage to become established in this managed environment routinely are mowed or sprayed.

Burgeoning population growth in the basin compelled the Department of Water and Power to extend a system of canals and aqueducts far beyond the Los Angeles basin. These tentacles now sprawl from the Colorado River up the eastern Sierra Nevada to the Mono Basin, diverting water from large and small streams that cannot support riparian habitats without it. Whole reaches of the Owens River that once were mantled with luxuriant thickets of cottonwood and willow, today, are parched open scrubs of scattered sagebrush, rabbitbrush, or the invasive tamarisk. These thickets once rang with the calls of yellow billed cuckoos, Bell's vireos, and willow flycatchers, birds which are almost extinct today.

Surprisingly, this staggering habitat loss went unappreciated until the early 1970s. Federal and state wildlife agencies finally became concerned and initiated inventories of surviving riparian resources on several California stream systems. One such study on the Sacramento River documented a twenty percent reduction in acreage of privately owned riparian habitat from 1972 to 1977. Another study in Western San Diego County showed nearly a two-thirds reduction in the stream miles that supported riparian habitats between 1976 and 1987. Similar habitat losses have occurred on most low-elevation California stream systems. Given such extensive habitat depletion, it should come as no surprise that so many riparian dependent animals have found their way onto lists of threatened or endangered species.

What Has Happened to Cause Such Losses?

These losses have not been due to natural causes, but are due to human activities. Native American burning practices undoubtedly were applied to riparian settings. Many central coast corridors were extensively modified by the Indians, particularly in Santa Barbara and Ventura counties. The Piutes had extensive ditch-and-flood irrigation systems in the Owens Valley. The extensive distribution of buckeye in San Mateo and Santa Cruz County stream corridors has been attributed to deliberate establishment by local Costanoan Indians, who depended heavily on buckeye as a food source.

The first European activity to affect California riparian habitats undoubtedly was livestock grazing. The early vaqueros recognized that streams provided abundant water, palatable forage, and shade in an otherwise arid environment. Large areas around the distributary fans of Great Valley streams such as the Kaweah once supported open savannas of valley oak over perennial sod-forming grasses. These were extensively grazed in the years before flood control allowed cultivation to encroach upon the floodplain. Cattle and sheep either eat or trample the tender young shoots, leading to "browse lines" on the undersides of trees below which few leaves survive. Today, these are most evident in mountainous areas used for summer grazing or around water sources in the desert, where livestock tend to aggregate during daytime.

Marshall's discovery of gold on the American River near Coloma had myriad impacts on California's natural resources, including its riparian habitats. Much of the early mining occurred along streams; riparian vegetation was considered overburden that impeded gold recovery, so large tracts were cleared. The large amounts of water that hydraulic mining required prompted development of a vast system of stream diversions, canals, and flumes that dewatered streams for use in the mines. This dewatering had the same effect then as present day diversions for hydropower generation are having on montane streams throughout California. Many of the early ditch systems still divert water, but to hydroelectric plants and water supply agencies instead of mines. Agriculture uses more than eighty-five percent of California's water, much of it for irrigating the arid San Joaquin and Imperial valleys. This is water that otherwise would be available to support riparian growth.

The same hydraulic mining that dewatered streams also returned huge amounts of sediment to the state's waterways, clogging channels, changing flood patterns, and altering the conditions under which riparian habitats replace themselves.

Early Californians harvested large amounts of wood for heating or building. They had to depend entirely on wood for fuel until coal mines were developed late in the nineteenth century. Steam-powered paddle boats plied the state's lakes and rivers by the 1860s, fueled by crews that would cut and stack mounds of cordwood at strategic locations along the waterways. This same wood was loaded on barges for transport to the fuel wood markets in town.

Although the woods of most riparian trees are not well suited for structural lumber, they were used for certain applications. Apparently, rounds cut from the bases of sycamore trees were the preferred raw material for making ox cart wheels in Spanish and Mexican California. This suggests a ready availability of trees in excess of five feet in diameter. Today one would be hard pressed to find even a few sycamores this size. Timbers hewn from cottonwood logs still are used for flooring in farm trucks and trailers today.

The early mining and irrigation diversions were eclipsed in the middle of the twentieth century by much more grandiose systems of water storage for power generation and flood control. These larger reservoirs had their own impacts on the state's dwindling riparian resources. Flood control obviously reduces peak floods, but it also changes the timing of such flooding, usually by beginning water releases earlier than otherwise would occur. Before flood control, most free-flowing California streams became virtually dry during late summer or fall. Now, late season flows typically are augmented by controlled releases from upstream storage. As a result, many formerly intermittent California streams now flow year-round, altering the vegetation.

Reservoirs do more than alter flow regimes. They do not just fill with water, but also gradually fill with sediment that once was carried downstream and deposited in alluvial floodplains. Water released from storage does not carry this material, so the water is capable of picking up new sediment from the streambed below the impoundment. This leads to a gradual removal of the finer bedload below the dam. The results are downcutting and residually coarser streambeds that are more deeply incised in their channels and are less able to migrate laterally. Reduced lateral migration changes the types of sites available for plant regeneration, thereby initiating a change in species composition of the streamside community.

Groundwater pumping also affects riparian vegetation through its alteration of local water supplies. Sustained pumping has lowered water well levels all over California. Riparian plants that require plentiful and continuous water supplies may not be able to grow their roots deep enough to tap groundwater supplies under such pumping regimes. Losses in riparian cover along the Salinas and Carmel rivers have been attributed to groundwater withdrawals.

Flood control and its effect on water delivery schedules have resulted in substantial changes on several California streams. Flood protection allows encroachment into formerly flood-prone areas, and the conversion of natural habitats into urban development. Unfortunately, streamside soils are among the state's most productive. Flood protection and irrigation development have allowed vast tracts to be cleared for cultivation. This process has continued up to the present day; most of the losses in riparian habitat documented along the Sacramento River since 1972 have been the result of clearing for orchard development. Augmenting late summer flows on flood-controlled streams, however, has resulted in minor increases in riparian vegetation in many parts of California.

Wastewater treatment facilities sometimes play important roles in maintaining stream flows through the summer. A good example may be seen along the Santa Ana River between Riverside and Prado Dam. As it leaves Riverside, the Santa Ana is a bare streambed completely constrained by concrete or riprap. In the next dozen or so miles, several outfalls of secondary treated effluent augment stream flow. The otherwise seasonal Santa Ana becomes perennial, allowing a lush riparian corridor to prosper.

What is the Future for Riparian Habitat?

Several private organizations and public agencies have recognized the precarious status of riparian habitats and have risen to their defense. These include private groups concerned with fisheries, waterfowl, wildlife, recreation, and conservation.

State and federal agencies, such as the California Department of Fish and Game or the US Fish and Wildlife Service, have conducted inventories of surviving riparian resources, sometimes as a result of political maneuvering and demand by private groups. Most of these inventories have focused on the Great Valley or one of its natural subdivisions; at least four separate studies have been conducted in the Sacramento Valley alone. Knowledge of the status of coastal riparian habitats is less complete, but is far ahead of what we know about desert riparian resources.

Some sites have been acquired expressly for their riparian features. The California Department of Parks and Recreation operates Caswell Memorial State Park and Woodson Bridge State Recreation Area, both of which protect fairly large riparian remnants. The California Department of Fish and Game has acquired several parcels that include riparian habitats. Many of these were obtained to provide hunting or fishing opportunities to the general public, but some have been obtained strictly to protect riparian resources. The Nature Conservancy protects riparian resources at their Cosumnes River, Oxbow, and South Fork Kern River preserves. The Sacramento Audubon Society owns the Bobelaine Preserve on the lower Feather River. This land was willed to Audubon by a rancher who preferred to see his little patch of the Great Valley's jungle persist in perpetuity rather than be converted to cropland. Grass roots organizations have scored some impressive victories in riparian habitat protection, the most successful of which may be the Save the American River Association, which has secured protection of thousands of acres along the lower American River in Sacramento County. The American River Parkway is the only urban stream in the nation to be included in the original National Registry of Wild and Scenic Rivers.

Groups such as these have recognized that urban streams can be more than concrete-lined canals. Even urban streams can provide amenities such as recreation, wildlife viewing, or open space. This recognition has led to the embryonic field of riparian restoration: putting back the biota that had been removed by previous stream "improvements". The same techniques also are being used to restore riparian habitats in rural areas. The Nature Conservancy's efforts along the Cosumnes River in Sacramento County may be the largest wildland restoration project ever undertaken. A consortium of federal and state wildlife agencies has initiated a similarly ambitious project along the lower Colorado River. These and other restoration projects may provide suitable habitat for several riparian birds that have become endangered because of past habitat losses. Hopefully the projects will succeed before these species of birds are lost.

Pressures on riparian habitats will not diminish in the future. Multiple and conflicting demands for clean domestic, industrial, and agricultural water supplies, hydropower generation, and effluent dilution and transport all compete for the water needed to sustain riparian ecosystems. Those who are interested in assuring that their successors will enjoy these habitats must learn about the physical and biological processes that shape them, for it is only through informed, cogent arguments based on such facts that additional examples of this remarkable resource will persist.

ABOUT

The Field Guide

The purpose of this field guide is to provide laymen with an easy method for identifying common riparian plants in California.

Illustrations are direct photocopies of living plants (1:1 scale); therefore, the user can compare actual plants found in the field with illustrations in the book, taking into account that there are variations in growing conditions that result in size differences. Floral details are lost in this process but much of the gestalt used in plant identification remains. This field guide is thus developed for the layman to use in the field for the identification of the most common riparian plants.

Plants have been included in the guide from montane regions of the Sierra Nevada and the Coast Ranges and their foothills, and from the Great Valley, the desert, and coastal California. The list of plants included in the guide has been developed from the suggestions of several botanists; it is not an "official" list but includes what are generally considered to be common plants in the riparian community of California. Species are included that in the drier part of their range are restricted solely to a riparian corridor but, further to the north, are not so restricted. Plants have been collected during different parts of the season. In some cases species are illustrated by plants having juvenile leaves and flower buds, while in other cases, plants have mature flowers or even fruits. The leaf size of a plant collected in Northern California may be significantly larger than a leaf of the same species collected in drier regions to the south; and for very large plants, the leaf that accomodates itself to an 8 1/2 x 11 page is likely to be a juvenile leaf as in the prints of *Aralia californica* and *Washingtonia filifera*. Several exotic species are included, particularly those that are aggressively invasive weeds.

Plants are arranged alphabetically by genus and species name; thus, the plants of any particular genus are found together. For example, all the willows, members of the genus *Salix*, are together regardless of their life form, shrub or tree. Several genera such as *Salix, Rubus, Ribes*, and *Saxifraga* are so well represented in the riparian community that more than one example is included. Often additional species are listed below an illustrated species, with range and species differences provided. Because of the frequent disturbance innate to riparian habitats, there are few rare or endangered plant species. None are included in this guide.

Many plants have no common name and others have several. Where there is a common name, the generally accepted common name is provided in each description. Every plant has a single accepted scientific genus and species name designated by one author. We have followed the published names in Munz and Keck's *A California Flora* or those to be published in the new *Jepson Manual*, scheduled for publication in 1992.

Descriptions include information about a plant's habitat and the range in which it is found. If a plant looks similar to one in the book but has been found growing outside the described range, chances are it may be a plant in the same genus but a different species. Plant size, life form (shrub, tree, or herb), leaf shape, and flower and fruit information are all provided as additional guides to making a correct identification.

Figure 4. Map of California showing county boundaries.

PEA FAMILY *Fabaceae*

Catclaw

Acacia greggii

- Common in washes and canyons below 6000 feet; Creosote Bush Scrub to Pinyon Juniper Woodland in the Colorado and Mojave deserts.
- Spreading, deciduous, straggling shrub, branches armed with short, stout, recurved spines, 3 to 6 feet tall.
- Leaves divided into 4 to 6 pair of oblong leaflets.
- Yellow flowers forming cylindrical spikes, developing into compressed pods, constricted between the seeds; blooms April to June.

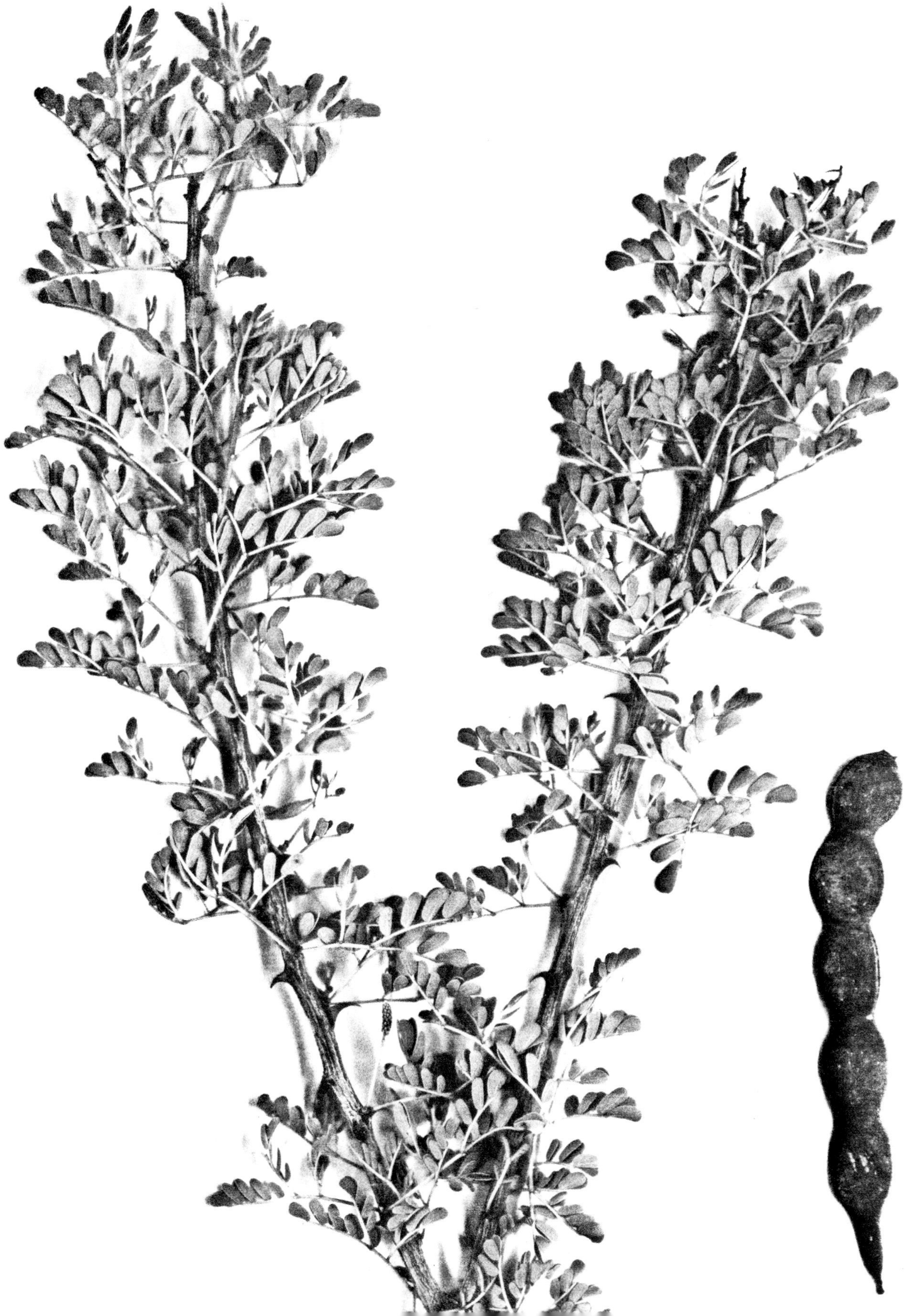

MAPLE FAMILY *Aceraceae*

Vine Maple

Acer circinatum

- Shaded stream banks below 5000 feet; Mixed Evergreen, Redwood, and Yellow Pine forests; Mendocino, Yuba, and Butte counties north.
- Deciduous shrub or small tree, 3 to 20 feet tall, often vinelike and reclining.
- Leaves thin and palmate, paler below.
- Relatively inconspicuous flowers, 4 to 10 in a cluster, developing into winged fruits; blooms April to May.

MAPLE FAMILY *Aceraceae*

Big-Leaf Maple

Acer macrophyllum

- Common on stream banks and in moist shady canyons below 5000 feet; most of California west of the Sierra Nevada.
- Round-topped deciduous tree, 15 to 100 feet tall.
- Leaves 3 to 5 parted with deeply incised and irregularly toothed lobes, paler and more downy beneath.
- Inconspicuous greenish flowers developing into winged fruits; blooms April to May.

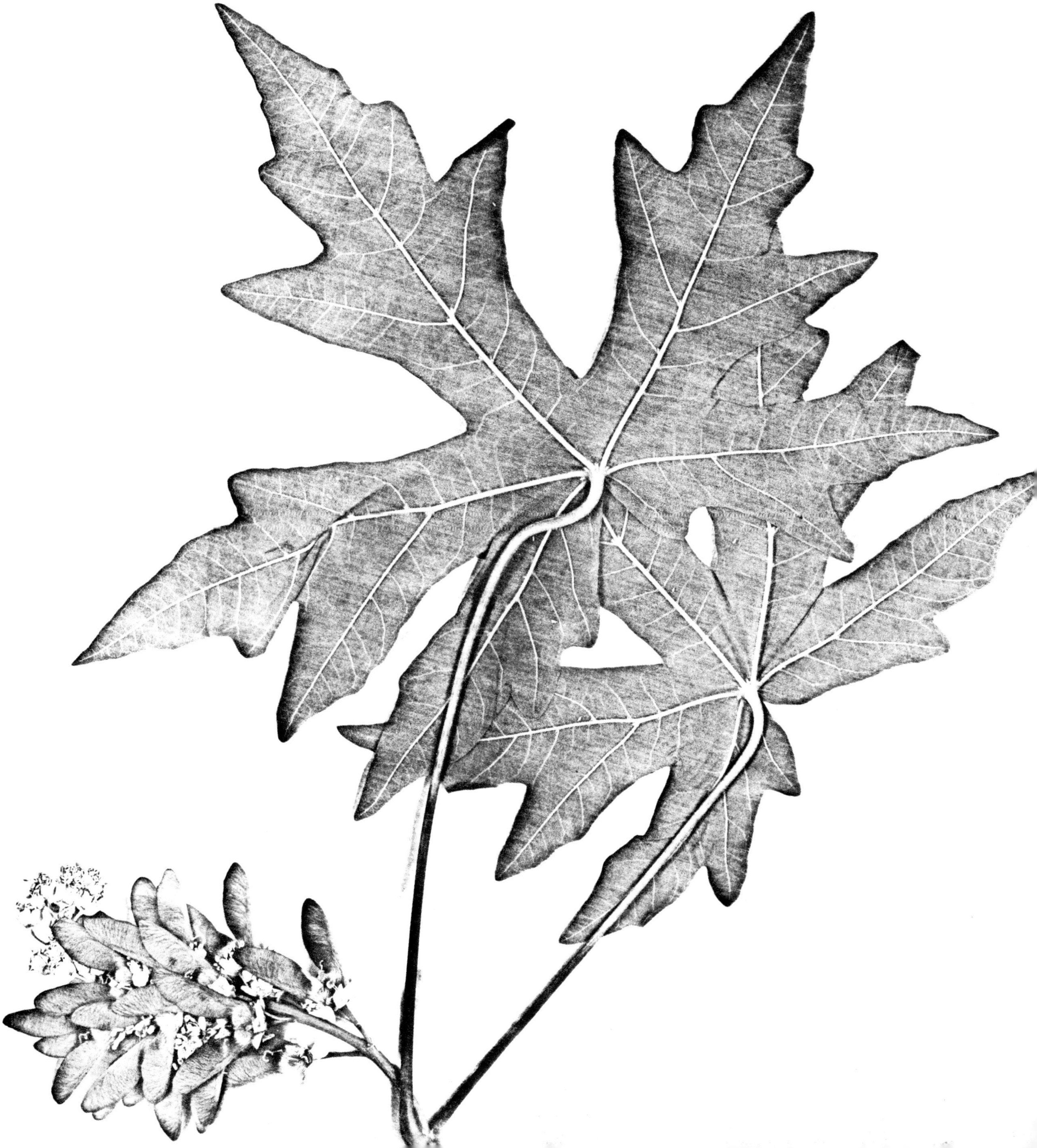

MAPLE FAMILY *Aceraceae*

Box Elder

Acer negundo
subsp. *californicum*

- Along streams and bottom lands below 6000 feet; west of the Sierra Nevada, from San Jacinto and San Bernardino mountains, Santa Barbara and Kern counties north to Shasta and Siskiyou counties.
- Broad, open-crowned, deciduous tree, 20-50 feet tall, often with several stems, separate male or female trees.
- Leaves 3-parted with coarsely-toothed leaflets, woolly beneath; the only compound-leaf maple in North America.
- Female trees have small, greenish flowers that develop into winged fruits (a double samara); male trees have clusters of drooping threadlike flowers; blooms March to April.

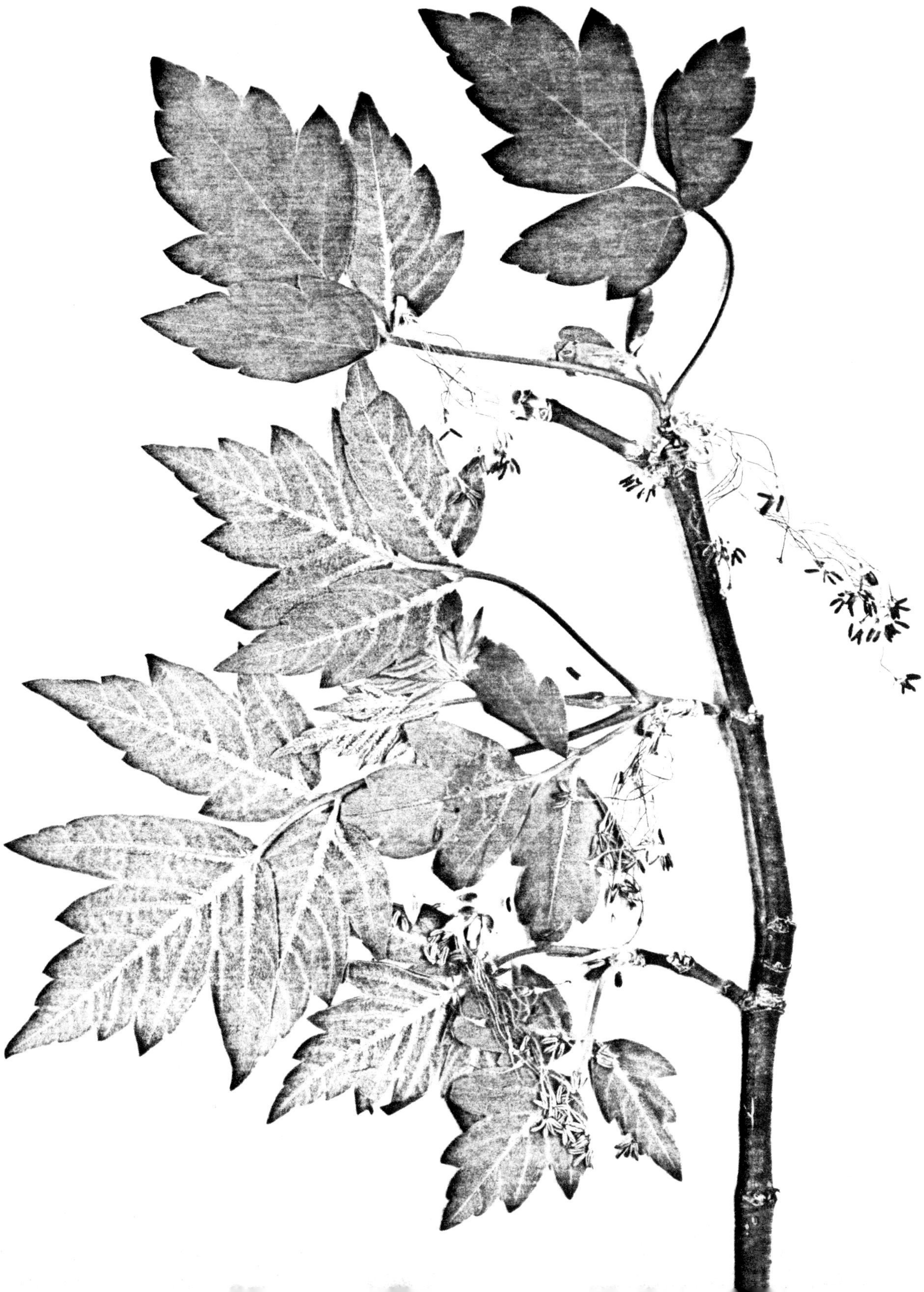

BUTTERCUP FAMILY *Ranunculaceae*

Monkshood

Aconitum columbianum

- Moist places such as meadows, especially around willow thickets, 4000 to 8000 feet; Montane Coniferous Forest of Sierra Nevada and north Coast Ranges south to Trinity County.
- Perennial herbs with stems erect, or weak and reclining, 20 to 80 inches tall.
- Leaves thin and deeply cleft.
- An elongate, showy flower cluster with purplish-blue, hooded flowers developing into 3-5 dry follicled fruit; blooms July through August.

FERN FAMILY *Aspidiaceae*

Five-Finger Fern

Adiantum pedatum

- Moist shaded rock crevices and in swampy woods and canyons from sea level to 10,700 feet; Redwood, Mixed Evergreen, and Montane Coniferous forests; Coast Ranges and Sierra Nevada.
- Delicate fern with short rhizome, fronds 8 to 32 inches wide and long, divided at base and into branches resembling fingers
- Reproduction by spores from sori on underside near margins of fronds.

BUCKEYE FAMILY *Hippocastanaceae*

Buckeye

Aesculus californica

- Common in canyons and on dry slopes, below 4000 feet in foothill woodlands of the Coast Ranges and Sierra Nevada from Los Angeles north to Siskiyou and Shasta counties.
- Large deciduous bush or tree with broad round top, to 21 feet tall.
- Palmate leaves with 5-7 leaflets.
- Flowers in long inflorescence, white fading to pink and developing into a fruit with a leathery capsule and large shiny seed; blooms May-June.

Indians ground up seed to use for stupefying fish; flower pollen is poisonous to bees.

BIRCH FAMILY *Betulaceae*

Red Alder

Alnus oregana

- Stream banks and marshy places below 500 feet; Mixed Evergreen and Redwood forests; Santa Cruz County north to Oregon.
- Deciduous tree to 75 feet, often reduced to shrubs in exposed places.
- Leaves coarsely and variously toothed, broadly ovate, pointed at tip and rounded at base, obscurely-rolled leaf edge.
- Male catkins clustered at ends of branchlets, female flowers in roundish cones; blooms March to April.

Alders are notable for their ability to fix nitrogen in root nodules, formed in response to a symbiotic actinomycete fungus that can assimilate atmospheric nitrogen.

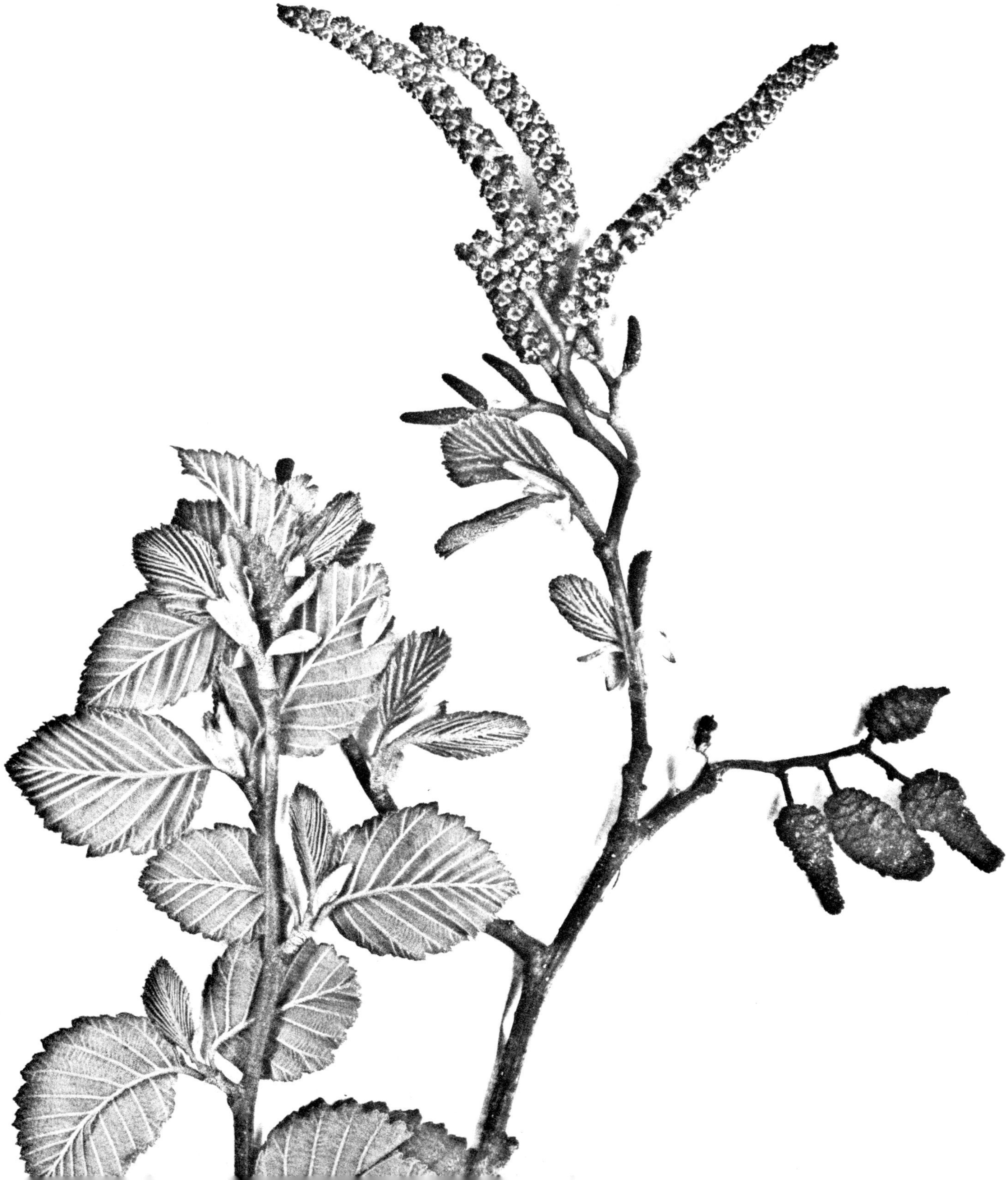

BIRCH FAMILY *Betulaceae*

White Alder

Alnus rhombifolia

- Along boulder-strewn streambanks, mostly below 5000 feet; much of California west of the Sierra Nevada except immediate strip along North Coast where it is replaced by red alder.
- Deciduous tree 30 to 100 feet tall with whitish or greyish-brown bark.
- Leaves similar to red alder without rolled edge.
- Male catkins and female cones similar to red alder.

ROSE FAMILY *Rosaceae*

Service-Berry

Amelanchier pallida

- Dry gravelly and rocky slopes and flats, below 11,000 feet; many plant communities but especially in the Montane Coniferous Forest of the Coast Ranges and Sierra Nevada; San Diego County north.
- Deciduous shrub 3 to 9 feet, with rigid erect to spreading branches and bark reddish-brown becoming grey shredded-like.
- Simple oval leaves, toothed at leaf tip down to middle.
- White flowers developing into dark-purple pome fruit; blooms May to June.

Service-berries provide important food supplies to several wildlife species during spring.

DOGBANE FAMILY *Apocynaceae*

Indian Hemp

Apocynum cannibinum var. *glaberrimum*

- Occasional in moist places below 5000 feet through most of California.
- Perennial herbs with upright branching stems, milky juice; 12 to 24 inches tall.
- Leaves opposite, undivided.
- Flowers small pale in flat topped clusters, often twisted in bud, developing into fruit with 2 slender follicles; blooms June to August

BUTTERCUP FAMILY *Ranunculaceae*

Columbine

Aquilegia formosa var. *truncata*

- Common in moist places below 8000 feet; most plant communities, Sierra Nevada and Coast Ranges, San Diego County north.
- Perennial growing from thickened base; 20 to 40 inches tall.
- Cleft to the middle, lobed, and round toothed.
- Nodding with red sepals and spur, yellow petal parts, fruit of 5 erect follicles; blooms April to August.

Columbine *A. schockleyi* is found in springy places 5000 to 7700 feet in the Pinyon-Juniper Woodlands of the Panamint, Argus, Clark, and New York Mountains, Mojave Desert.

GINSENG FAMILY *Araliaceae*

California Aralia, Elk clover

Aralia californica

- In moist and shaded places below 5000 feet; many plant communities west of the Sierra Nevada, particularly in the central Coast Ranges growing with white alder; Orange County north.
- Large-leafed perennial, roots large with milky juice; 3 to 9 feet tall.
- Leaves large with 3 to 5 small-toothed leaflets.
- Small white flowers in a large inflorescence arranged in several small ball-like clusters, fruit berrylike; blooms June to August.

BIRTHWORT FAMILY *Aristolochiaceae*

Dutchman's Pipe, Pipevine

Aristolochia californica

- Stream banks growing over shrubs, below 1500 feet; Foothill Woodland, Mixed Evergreen Forest in the Coast Ranges from Monterey County north, Sierra Nevada foothills from Sacramento County north.
- Twining or climbing perennial vine, 9 to 12 feet long.
- Oval heart-shaped thin leaves.
- Two-lipped, purple-veined, curved, tubular, pipelike, flower; fruit a many-seeded capsule; blooms January to April.

The pipevine plant synthesizes glycosides which are capable of inducing heart attacks in vertebrates. Larvae of the pipevine swallowtail butterfly feed on the leaves, lacing their bodies with glycosides as a defense against predators. Several other butterfly species mimic the coloration of the pipevine, their defense is so successful and the predators so wary.

SUNFLOWER FAMILY *Asteraceae*

Mugwort

Artemesia douglasiana

- Moist low places, below 6000 feet; many plant communities west of Sierra Nevada.
- Perennial herb; 20 to 60 inches tall.
- Leaves alternate, coarsely toothed or lobed, usually aromatic, densely grey-hairy on underside.
- Flowers small, greenish in a dense elongate cluster, fruit small dry akenes; blooms June to October.

GRASS FAMILY *Poaceae*

Giant Reed

Arundo donax

- Moist places along streams, seeps, and ditches; desert and west of the Sierra Nevada in California and abundant along leveed channels in the Sacramento-San Joaquin delta.
- An introduced perennial reed; 20 to 23 feet tall.
- Leaf blades to 1-1/2 inches wide.
- A large, plume-like, terminal inflorescence; blooms March through September.

This plant is aggressively invading stream corridors in much of Southern California. Flood events seem to favor this species over several native species.

FERN FAMILY *Aspidiaceae*

Lady Fern

Athrium felix-femina
var. *sitchense*

- Along streams and in swampy places at low elevations near the coast; Mixed Evergreen and Redwood forests; San Luis Obispo County north to Oregon.
- Medium sized to large upright fern.
- Fronds generally large, blades 2 to 3 times pinnate.
- Sori on under surface, moon-shaped or oblong, with toothed indusia.

SUNFLOWER FAMILY *Asteraceae*

Mule Fat

Baccharis viminea

- Dry stream beds and ditch banks of Valley Grassland, Foothill Woodland, and Coastal Sage Scrub, mostly below 1500 feet; Sacramento Valley and bordering foothills through the Coast Ranges, San Diego County north to Tehama and Butte counties.
- Willow-like woody shrub, 6 to 12 feet tall.
- Leaves small, mostly entire, some finely-toothed.
- Male and female flowers on separate plants; whitish inconspicuous flowers in small clusters at tips of short lateral branches, fruit of achenes bearing copious fine bristles; blooms March through July, some year-round.

Baccharis viminea and *B. glutinosa* are combined into the latter in some taxonomic treatments. Seep Willow *B. glutinosa* is also common, forming thickets along watercourses up to 2500 feet in several communities in the Owens Valley through the Mojave and Colorado deserts to San Diego County. Flowers are in terminal clusters; blooms April to October.

BIRCH FAMILY *Betulaceae*

Resin Birch

Betula glandulosa

- Wet places 6500 to 7500 feet; Lodgepole Forest from Mount Lassen to Modoc County.
- Shrub, mostly 3 to 6 feet tall, with warty glandular twigs.
- Ovate to round leaves with rounded teeth.
- Male flowers in elongate catkins; female in erect catkins; blooms April to June.

Water birch *Betula occidentalis* occurs in moist places 2000 to 8000 feet, mostly in the Montane Conferous Forest of Inyo and Tulare to Modoc and Humboldt counties.

FERN FAMILY *Blechnaceae*

Deer Fern

Blechnum spicant

- Wet sheltered places; Redwood and Mixed Evergreen forests; Santa Cruz to Del Norte Counties.
- Ascending to erect fern, 4 to 32 inches tall, with short creeping rhizomes.
- Many evergreen sterile fronds, few fertile fronds that are larger and crowded on back with sporangia.

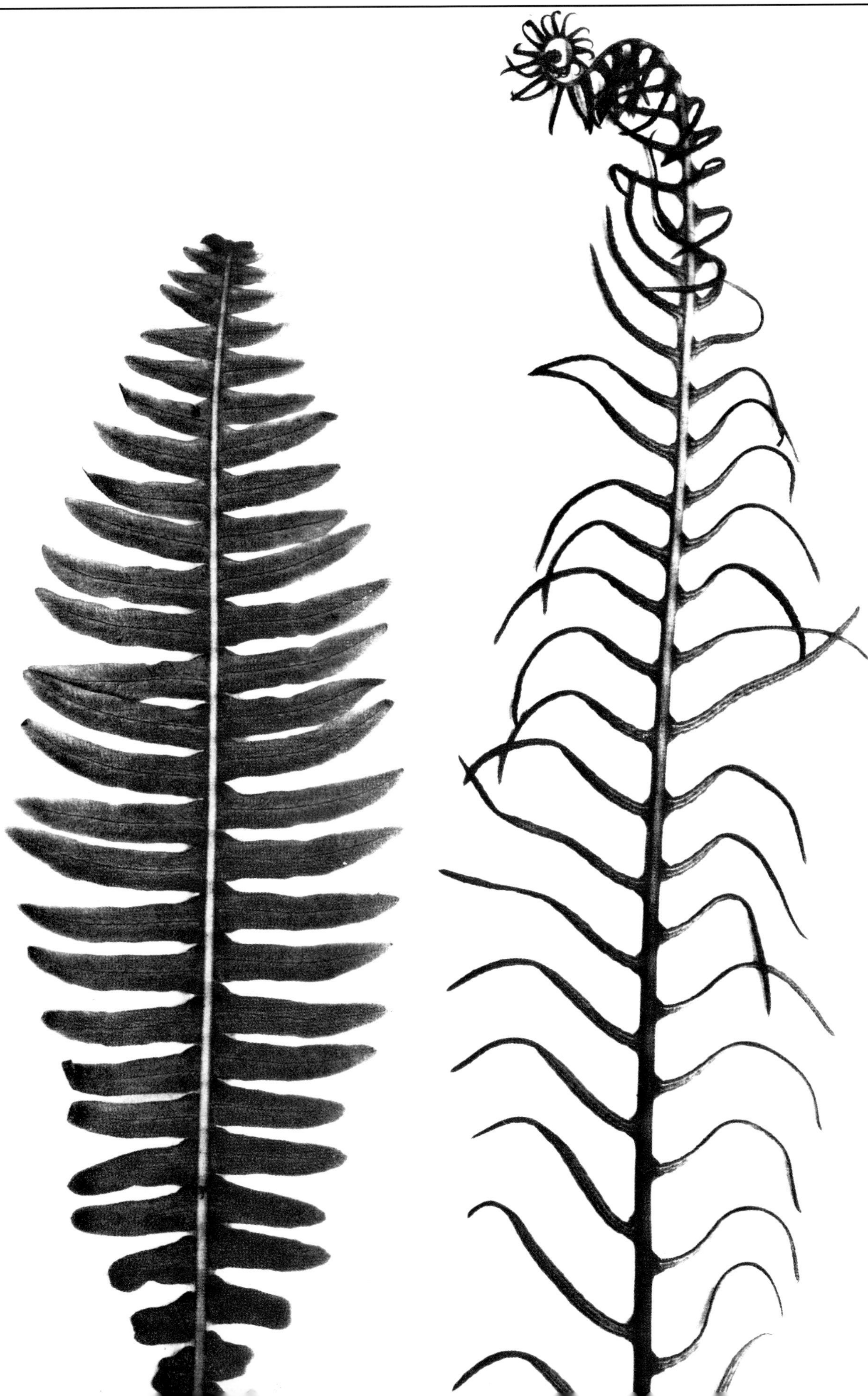

SAXIFRAGE FAMILY *Saxifragaceae*

Boykenia

Boykenia elata

- Shaded springy places below 5000 feet; Coastal Scrub, Mixed Evergreen, Redwood, and Yellow Pine forests; Coast Ranges from Santa Monica Mountains north to Del Norte County and the Sierra Nevada north from Amador County.
- Perennial herb, 8 to 24 inches tall, with a basal tuft of leaves, upper leaves reduced.
- Leaves thin and ovate with bristle pointed teeth.
- Many tiny whitish-flowered elongate inflorescence, fruit a 2-beaked capsule with many small seeds; blooms June to July.

CALYCANTHUS FAMILY
Calycanthaceae

Spice-Bush

Calycanthus occidentalis

- Moist places, canyon streams and about ponds below 4000 feet; Foothill Woodland and Yellow Pine Forest of the northern Coast Ranges and west base of Sierra Nevada, Tulare to Shasta counties.
- Erect, rounded, deciduous shrub 3-9 feet tall, pleasantly aromatic when bruised.
- Leaves oval to oblong, somewhat hairy beneath.
- Flowers solitary, terminal, and a deep reddish brown developing into an oval cuplike fruit; blooms April to August.

MUSTARD FAMILY *Brassicaceae*

Milk-Maid, Toothwort

Cardamine californica

- Shady banks and slopes mostly below 2500 feet; many plant communities from San Diego County north, mostly through Coast Ranges, occasional in Sierra Nevada.
- Perennial growing from deep rhizomes; 4 to 16 inches tall.
- Rhizome leaves divided into three ovate leaflets, stem leaves divided into 3 to 5 toothed, ovate, or lance-shaped leaflets.
- Pale-rose to white flowers with 4 petals in clusters, flattened, linear seed-pods that open elastically; blooms February to May.

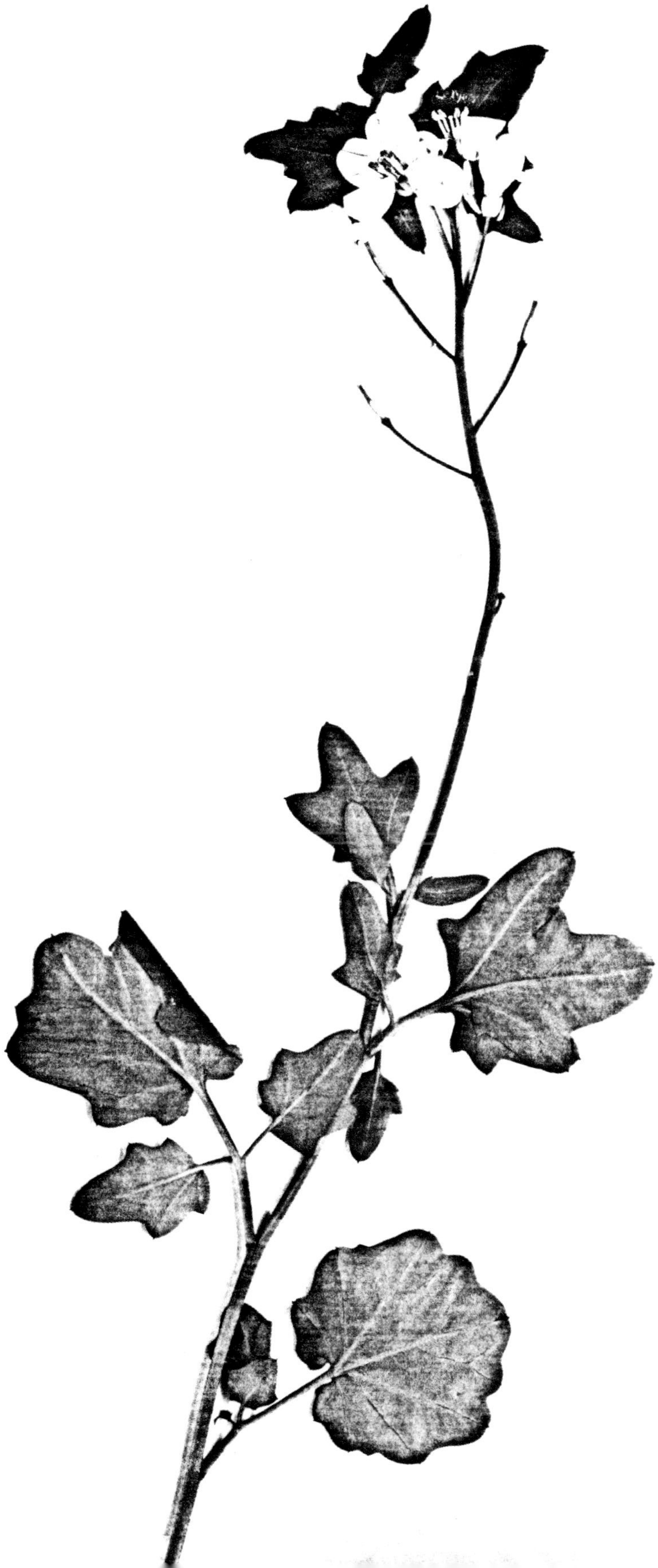

SEDGE FAMILY *Cyperaceae*

Barbara Sedge

Carex barbarae

- Open or brushy slopes of valley flats that are wet in spring, to 3000 feet; Valley Grassland, Foothill Woodland, and Coastal Prairie to Mixed Evergreen and Yellow Pine forests; coastal Southern California north through Coast Ranges and occasional in Sierra Nevada foothills.
- Perennial growing in large clumps, 12 to 40 inches tall.
- leaf blades to 1/3 inch wide with a lower sheath.
- Male and female flower spikes separate with one or two male spikelets above the female spikelets, the ovary borne in a flask-like bract.

The genus *Carex* in the Sedge Family is one of the largest genera of plants in the world with more than 1000 species found world-wide, 144 species found in California, and 45 in the wetlands of California. Identification generally requires microscopic examination of flowering parts. The California Indians cultivated this species, *C. barbarae*, for use in basketry.

MADDER FAMILY *Rubiaceae*

Buttonbush, Button Willow

Cephalanthus occidentalis var. *californicus*

- Along lake shores and streams below 3000 feet; Foothill Woodland and Valley Grassland; Central Valley and bordering ranges, Kern to Siskiyou counties.
- Shrubs or small trees, 6 to 27 feet tall, with grey bark or reddish twigs.
- Leaves opposite, long-oblong, smooth, and pointed at tip.
- Small white flowers in heads, clustered into round heads on long flower stalks, fruits small and dry; blooms June through September

PEA FAMILY *Leguminosae*

Palo Verde

Cercidium floridum

- Washes and low sandy places below 1200 feet; Creosote Bush Scrub in the Colorado Desert.
- Summer-deciduous tree, up to 30 feet tall, with blue-green trunk and spiny twigs.
- Leafless most of the year, leaves feathery with opposite pinnate leaflets.
- Flowers pea-like, yellow, in slender clusters in axils of leaves; fruit in long flat pods constricted between seeds; blooms March to May.

BIGNONIA FAMILY *Bignoniaceae*

Desert-Willow

Chilopsis linearis

- Common along washes and watercourses below 5000 feet; Creosote Bush Scrub and Joshua Tree Woodlands in the Mojave and Colorado deserts.
- A large, deciduous, few- to many- stemmed shrub, 6 to 18 feet tall.
- Leaves simple, entire, narrow, and alternate.
- Large, showy, funnel-shaped, lavendar or pink to whitish flowers, developing into very long, thin, flat pods; blooms May to September.

BUTTERCUP FAMILY *Ranunculaceae*

Clematis, Virgin's Bower

Clematis ligusticifolia

- Along streams and in moist places below 7000; many plant communities in Sierra Nevada and Coast Ranges.
- Semi-woody vine that climbs over bushes and trees by twining tendrils.
- Compound leaves with 5-7 leaflets.
- Flowers white woolly-silky, inside and out, few to many clustered in head; many small one-seeded fruits with long feathery-plumose styles; blooms March to August.

Infusions of this vine were used by early settlers for sores and cuts on horses, and by Indians for colds and sore throats.

DOGWOOD FAMILY *Cornaceae*

American Dogwood

Cornus stolonifera

- Moist places below 9000 feet, mostly in Montane Coniferous Forest; Southern Sierra Nevada to Siskiyou, Del Norte, and Trinity counties, and along sloughs and quiet waterways in the Delta region.
- Spreading shrubs 6 to 30 feet tall, often rooting at tips of branches, red-purple twigs.
- Opposite leaves with 4-7 prominent veins that form thin latex threads when broken.
- Many inconspicuous flowers in flat-topped clusters appearing with or after the leaves, developing into small round whitish stone fruits; blooms May to July.

C. sessilis, occurring along stream banks, 500 to 5000 feet in the mid-Sierra Nevada north, and the northern Coast Ranges, has small white flowers in heads subtended by brownish-yellow petal- like bracts; Brown dogwood *C. glabrata* occurring in moist places below 5000 feet in many plant communities, has 3-4 veins on each side of the leaf mid-rib; Mountain dogwood *C. nuttalii*, occurring below 6000 feet in the Coast Ranges and Sierra Nevada, has large white to pinkish petal-like bracts maturing into a cluster of red stone fruits.

BIRCH FAMILY *Betulaceae*

Hazelnut

Corylus cornuta var. *californica*

- Damp slopes and shady stream banks below 7000 feet; many plant communities; Coast Ranges from Santa Cruz north, Sierra Nevada from Tulare County north.
- Open, spreading, deciduous shrub, 6 to 18 feet tall.
- Leaves rounded, doubly toothed, with soft pubescent hairs above and below.
- Female flowers (showing in image) small, red, tongue-like at ends of leafy shoots and appear before leaves emerge, male flowers on small catkins, smooth hazelnut fruit hard, but delicious, enclosed in 2 large united bracts; blooms January to April.

ROSE FAMILY *Rosacea*

Hawthorn

Crataegus douglasii

- Near streams and meadows, 2500-5500 feet; Foothill and Northern Oak woodlands, Mixed Evergreen, Yellow Pine, and Red Fir forests from Marin County north.
- Much-branched deciduous shrub or small tree armed with sharp thorns, 6 to 30 feet tall.
- Doubly-toothed, lobed, rounded leaves.
- White, sometimes pink, cup-shaped flowers in terminal inflorescence, fruit a purplish-black pome with bony nutlets; blooms May-June.

SAXIFRAGE FAMILY *Saxifragaceae*

Indian Rhubarb

Darmera peltata

- Banks of mountain streams below 6000 feet; Mixed Evergreen, Yellow Pine Forest; Sierra Nevada from Tulare to Siskiyou and Humboldt counties.
- Stout, stemless, perennial herb with thick fleshy rhizome, 12 to 40 inches tall.
- Large, broad, many-lobed, basal leaves with stem attachment on the underside (peltate).
- Small, white to pink, flowers in a long inflorescence, fruit a reddish, dry follicle; blooms April to July.

DATISCA FAMILY *Datiscaceae*

Datisca

Datisca glomerata . . . .

- Dry streambeds and washes in mountains and hills below 6500 feet; Yellow Pine and Mixed Evergreen forests, Foothill Woodland, and Chaparral from San Diego to Shasta County.
- Stout perennial herb, up to 3 feet tall.
- Sharply-incised, toothed, lance-shaped leaves.
- Separate male and female plants, flowers without petals, several in each axil of a leafy raceme; fruit a many seeded 1-celled capsule; blooms May to July.

GRASS FAMILY *Poaceae*

Hairgrass

Deschampsia caespitosa

- Wet meadows 6000 to 8500 feet in Southern California, 3300 to 12,500 feet in Sierra Nevada; Montane Coniferous Forest and Alpine Fell-fields; San Bernardino, Sierra Nevada, North Coast Ranges.
- Densely tufted perennial grass, 24-48 inches tall.
- Leaves mostly basal, flat or folded.
- An open nodding panicle of flowers; blooms July to August.

FUMITORY FAMILY *Fumariaceae*

Bleeding Heart

Dicentra formosa

- Damp shaded places below 7000 feet; Redwood, Yellow Pine, and Red Fir forests, Northern Oak Woodland in the Coast Ranges from Santa Cruz County north, the Sierra Nevada from Tulare County north.
- Perennial herb with basal leaves, or basal and stem leaves; 8 to 20 inches tall.
- Thin, much-dissected leaves.
- Several nodding, pink to whitish, flattened, heart-shaped flowers; fruit a dry capsule with shining black-crested seeds; blooms March to July.

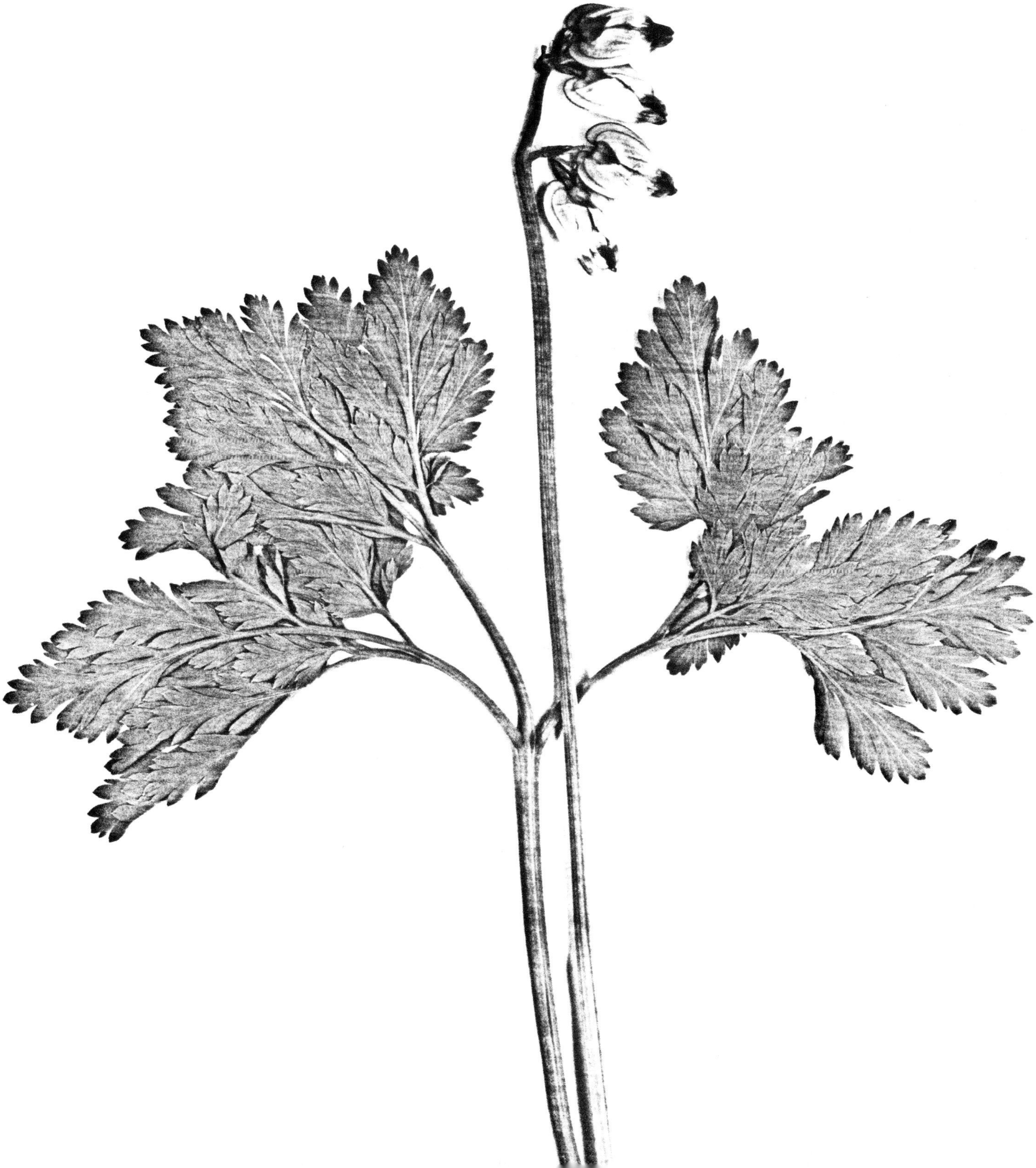

MEZEREUM FAMILY *Thymelaeceae*

Leatherwood

Dirca occidentalis

- Uncommon, wet slopes of rocky hills; Mixed Evergreen Forest in San Francisco Bay Area.
- Erect deciduous shrub with pliable twigs and smooth tough leathery bark; 24 to 80 inches tall.
- Oval leaves rounded at ends, smooth and light green above.
- Lemon-yellow flowers with long exserted stamen appearing before the leaves, fruit a yellowish-green stone fruit; blooms January to March.

LILY FAMILY *Liliaceae*

Fairy Bells

Disporum smithii

- Moist shaded woods near coast; Mixed Evergreen and Redwood forests; Santa Cruz County north to Oregon.
- Perennial herb with branched stems, 12 to 36 inches tall.
- Alternate oval leaves with 3-5 main parallel veins, pointed at tip.
- Drooping, whitish, cylindrical flowers in clusters of 1-5 developing into orange to red berries; blooms March to May.

GRASS FAMILY *Poaceae*

Creeping Wild-Rye

Elymus triticoides

- Moist and alkaline places below 7500 feet; many plant communities throughout the state.
- Sod-forming perennial with scaly creeping rhizomes, well-adapted to areas where fine silts or clays settle out of slow-moving floodwaters, 24 to 48 inches tall.
- Flat blades up to 1/3 inch wide.
- Awn-tipped flowers in 4-8 inch spikes; blooms June to July.

HORSETAIL FAMILY *Equisetacea*

Common Horsetail, Scouring Rush

Equisetum arvense

- Wet places below 10,000 feet; many plant communities throughout California west of the Sierra Nevada.
- Green rush-like perennial, erect, simple, or at length branched, rigid plant with creeping rhizomes, rough stems overlaid with silica, stomata arranged in rows in stem grooves; 8 to 24 inches tall.
- Whorled reduced leaves united into a sheath at node intervals.
- Pointed terminal cones producing microscopic spores.

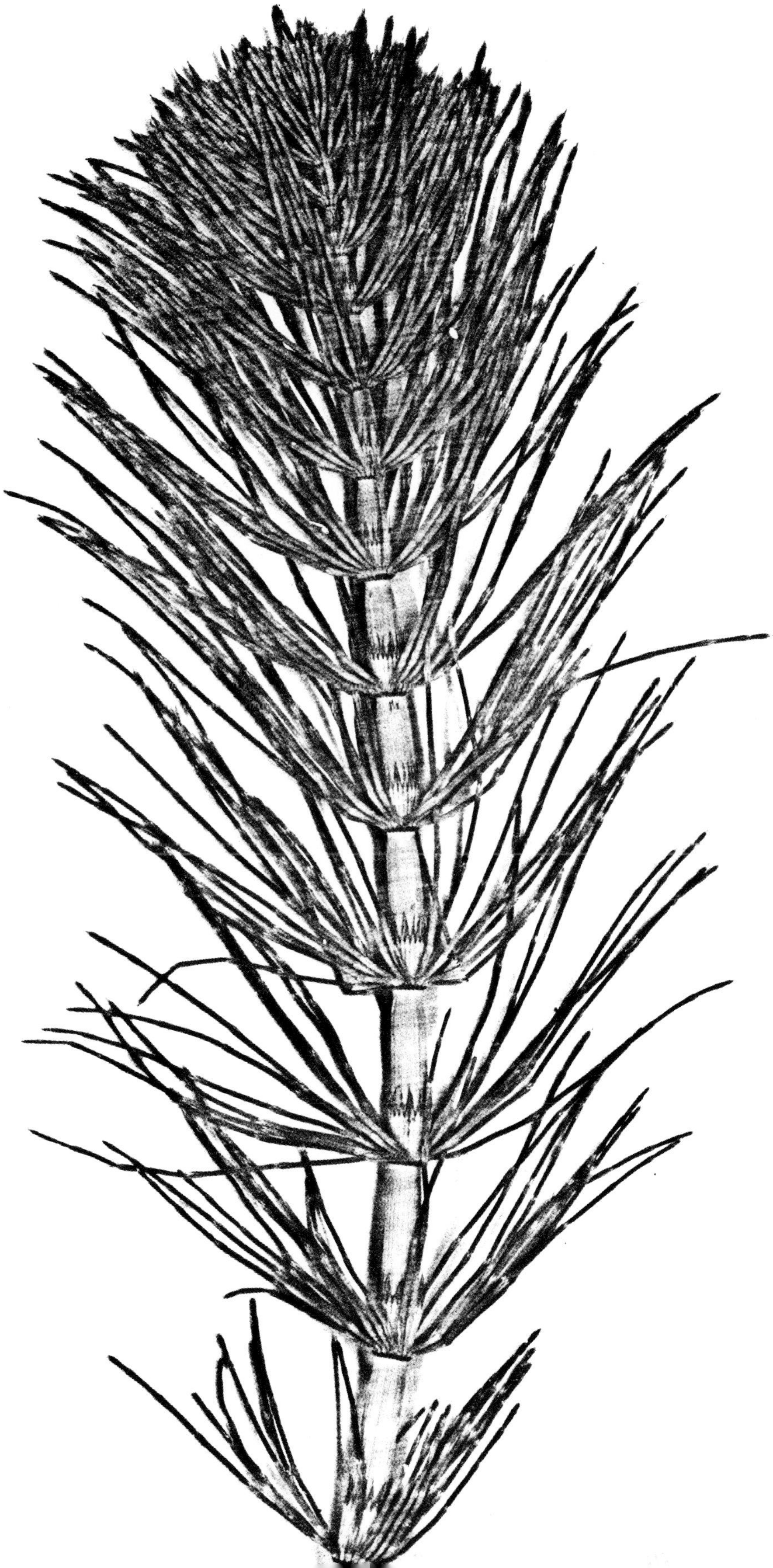

MULBERRY FAMILY *Moraceae*

Common Fig

Ficus carica

- An occasional, naturalized escape along shaded streams or flood plains, or along Delta levees.
- Much-branched deciduous tree to 30 feet.
- Leaves deeply 3-5 lobed and thick.
- Minute flowers inside a globose receptacle that ripens into a fleshy body, pollination internal by special insects; fruits in late summer.

OLIVE FAMILY *Oleaceae*

Oregon Ash

Fraxinus latifolia

- Canyons and near streams below 5500 feet; many plant communities, west base of Sierra Nevada from Kern County north to Modoc County, Coast Range from Santa Clara north.
- Deciduous deep-rooted trees, 30 to 75 feet tall, with sexes separate on different plants.
- Compound opposite leaves with 5-7 leaflets.
- Inconspicuous flowers in crowded clusters appearing before leaves, female flowers develop into clusters of 1-seeded winged fruit; blooms March to May.

Fraxinus latifolia is replaced by *F. velutina* in Southern California.

SUNFLOWER FAMILY *Asteraceae*

Sneezeweed

Helenium puberulum

- Moist meadows, marshes, and slough banks, mostly below 4200 feet in the Sierra foothills, frequent in the north Coast Ranges.
- Short-lived perennial blooming first year, 16 to 32 inches tall; single stems or few-branched.
- Leaves drooping, lower with leaf-stalks, upper generally without and smaller.
- Flowers in solitary, long-stemmed daisy-like heads, petal-like yellow ray flowers, central disk depressed-globose and sometimes red-purple, achene fruit topped with chaffy scales; blooms June to August.

Bigelow's sneezeweed *H. bigelovii*, common in moist meadowy places on both slopes of the Sierra Nevada, the mountains of Southern California, and the north Coast Ranges, has showy petal-like ray flowers; blooms June to August.

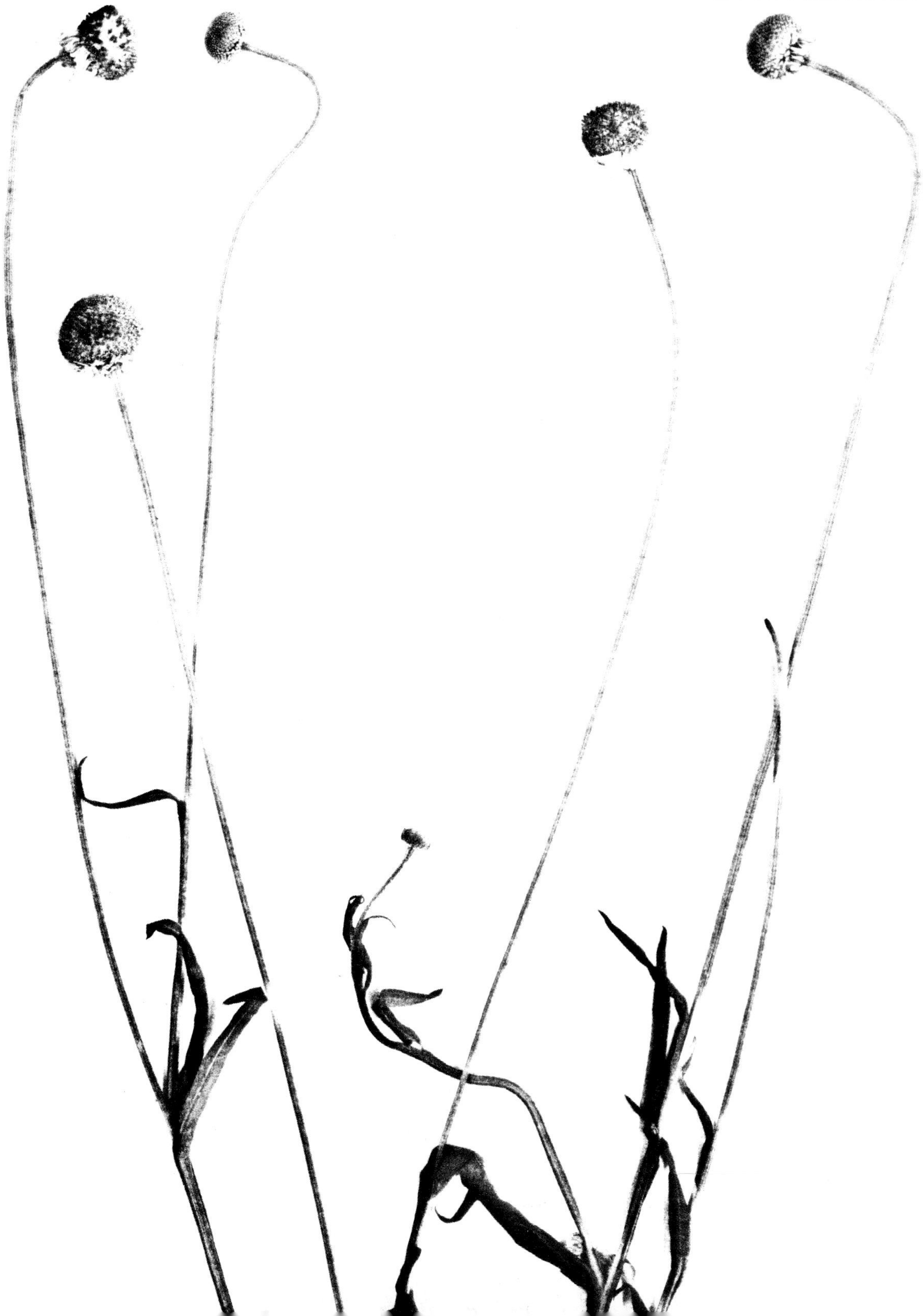

PARSLEY FAMILY *Apiaceae*

Cow Parsnip

Heracleum lanatum

- Moist shaded places below 9000 feet; many plant communities, San Jacinto and San Bernardino Mountains, Sierra Nevada and Coast Ranges north from Monterey County.
- Tall stout perennial, 3 to 9 feet.
- Large, lobed and toothed, broad, compound leaves.
- Clusters of small white flowers in a large loose umbel, developing into flattened thin-winged fruits; blooms April to July.

ROSE FAMILY *Rosaceae*

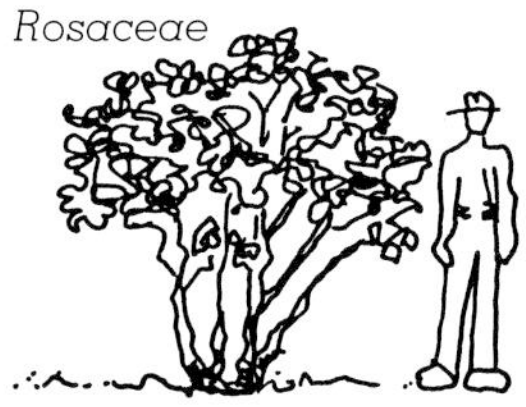

Toyon

Heteromeles arbutifolia

- Common on semi-dry brushy slopes and in canyons below 4000 feet; Chaparral and Foothill Woodlands in riparian setting in Southern California.
- Evergreen, freely-branching, tree-like shrub; 6 to 30 feet tall.
- Oblong, toothed, leathery leaves.
- Small white flowers in clusters developing into red berry-like fruit; blooms June to July.

SAXIFRAGE FAMILY *Saxifragaceae*

Alum-Root

Heuchera micrantha var. *pacifica*

- Moist banks of humus or rocks below 2000 feet; Closed-cone Pine, Mixed Evergreen, and Redwood forests; near coast from San Luis Obispo north to Oregon.
- Perennial herb.
- Roundish basal leaves with broad lobes and pointed teeth on long hairy stems.
- Small white flowers with petals narrowed at the base on a long flowering stem, capsule with 2 divergent beaks; blooms May to July.

MINT FAMILY *Labiatae*

Desert-Lavender

Hyptis emoryi

- Common in washes and canyons below 3000 feet; Creosote Bush Scrub in Colorado and Southern Mojave Desert.
- Erect, whitish, scaly, fragrant shrub with numerous slender straight branches, 3 to 9 feet tall.
- Leaves opposite, scalloped and toothed.
- Small violet flowers in clusters in leaf axils; blooms January to May.

WALNUT FAMILY *Juglandaceae*

California Black Walnut

Juglans hindsii

- Apparently this tree was once restricted to a handful of sites between the Bay Area and the Sacramento-San Joaquin Delta; native California Indians are known to have moved the species as far as Yosemite.
- Medium-sized deciduous tree; 45 to 75 feet tall.
- Long, pointed leaflets on compound leaves.
- Male flowers in pendant catkins; female flowers, few, at ends of new shoots maturing into a stone fruit within a fibrous, somewhat-fleshy, outer covering; blooms April and May.

Widely used as rootstock in commercial English walnut culture, this species and *J. nigra* (Eastern black walnut) and their hybrids are well established in mixed riparian forests in many parts of the Great Valley.

RUSH FAMILY *Juncaceae*

Baltic Rush

Juncus balticus

- Moist streamsides and backwaters, usually in sunny places, mostly below 5000 feet, throughout California.
- Wiry, round, grasslike perennial, 12 to 32 inches tall, not found growing in clumps.
- Flowers small with purplish-brown scale-like parts in clusters seeming to emerge from lateral portion of stem; fruit small, brown, ovoid capsules; blooms May to August.

Common rush *J. effusus* and its several varities, common in moist places below 8,000 feet throughout California, are wiry perennials growing to 4 feet tall in dense clumps of bright- greenish color; blooms June to August. The genus *Juncus*, commonly called rush or wire grass,is in a large group of over 200 species distributed world-wide, with 49 species found throughout California, many in riparian areas. Rushes are usually found in moist places growing with grasses and sedges.

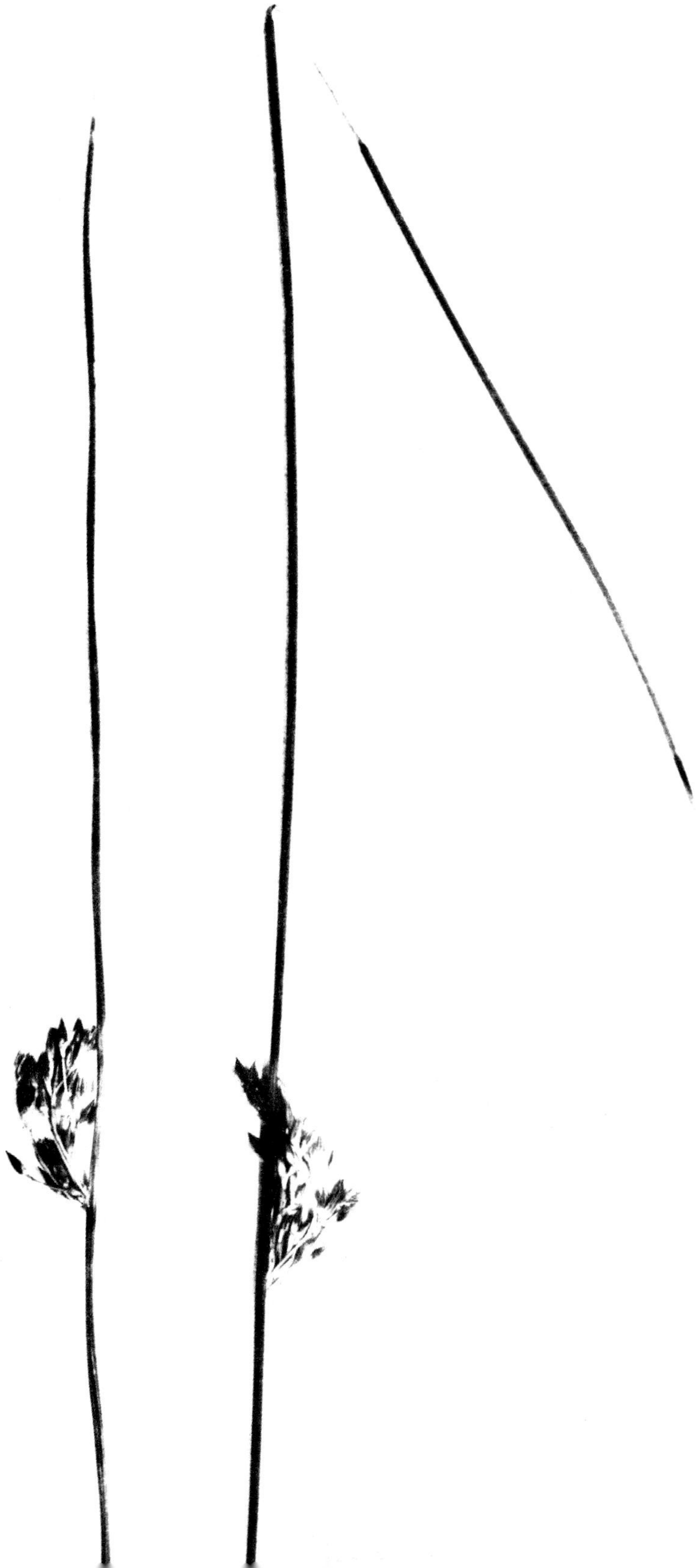

HONEYSUCKLE FAMILY *Caprifoliaceae*

Honeysuckle

Lonicera hispidula var. vacillans.

- Along streams and on wooded slopes below 2500 feet; Mixed Evergreen, Redwood, Douglas Fir forests, and Foothill Woodland; Sierra Nevada and Coast Ranges.
- Widespread, slender-stemmed, opposite-leaved, climbing vine, 6 to 18 feet tall.
- Leaves elliptic to oblong, the upper leaves usually united around the stem.
- Funnelform 2-lipped flowers, purplish or pink, in many whorls forming spikes; fruit a round red berry; blooms April to July.

HONEYSUCKLE FAMILY *Caprifoliaceae*

Twinberry

Lonicera involucrata var. ledebourii

- Moist places along California coast (var. *ledebourii*); moist places below 10,000 feet in Sierra Nevada.
- Erect or twining, deciduous shrub, 2 to 10 feet, montane form generally smaller.
- Leaves ovalish, opposite; the uppermost pair often joined around the stem.
- Yellow tubular flowers in pairs in leaf axils, subtended by showy bracts that often turn reddish or purple; fruit of paired, black, few-seeded berries set in colorful bracts; blooms March through April (coastal), June-August (montane).

Double-flowered honeysuckle *L. conjugialis* grows on wooded slopes, 4000 to 10,000 feet in the Sierra Nevada. A straggling deciduous shrub, it has small, paired, purple flowers, minute or lacking bracts, and its twin bright-red fruits are fused at maturity; blooms June to July.

PEA FAMILY *Fabaceae*

Lupine

Lupinus latifolius

- Common in open woods and thickets below 7000 feet, varieties up to 11,000 feet; many plant communities from coast to Red Fir Forest, Los Angeles to Del Norte and Siskiyou counties.
- Erect herbaceous perennial, 12 to 48 inches tall.
- Palmate leaves with 7-9 leaflets, 2 to 4 inches long, the leaf stem twice as long.
- Pea-shaped flowers on a long spike with blue or purplish petals fading to pink then brown; mottled seeds in brown, hairy, pea-like pods; blooms April to July.

ARUM FAMILY *Araceae*

Yellow Skunk-Cabbage

Lysichiton americanum

- Open swamps and wet woods near the coast; Northern Coastal Scrub and Redwood Forests; Santa Cruz to Del Norte Counties.
- Stemless ill-scented herb from a thick horizontal rootstock, leaves 12 to 60 inches.
- Leaves large, oblong to elliptic, basal, sheathing.
- Yellow to greenish-yellow broad sheathing flower (spathe) developing into green berry-like fruit; blooms April to June.

LILY FAMILY *Liliaceae*

False Lily-of-the-Valley

Maianthemum dilatatum

- Moist shaded banks near the coast; Redwood, Mixed Evergreen, and Closed-cone Pine forests; Coast Ranges, Marin County north.
- Low perennial herb with creeping rootstock, 6 to 14 inches tall.
- Broad heart-shaped leaves with parallel veins.
- Small white flowers in a terminal raceme, fruit a 1-2 seeded berry; blooms May to June.

GOURD FAMILY *Cucurbitaceae*

Wild Cucumber

Marah fabaceus

- Banks and slopes below 2500 feet; Coastal Strand, Mixed Evergreen Forest near coast; Marin to Monterey counties, common in the Sacramento Valley.
- Soft-stemmed climbing perennial, 9 to 21 feet long with lateral tendrils.
- Leaves broad, palmate, and 5-7 lobed.
- Cream colored flowers in leaf axils, clumped male and solitary female flowers separate but on the same plant; fruit a spiny gourd enclosing usually 4 large asymetrical flattened brownish seeds and tapering to a tip; blooms February to April.

FIGWORT FAMILY *Scrophulariaceae*

Scarlet Monkey-Flower.

Mimulus cardinalis

- Frequent on stream banks and seeps below 8000 feet; many plant communities in mountains and west of Sierra Nevada.
- Freely-branched perennial, 10 to 32 inches tall.
- Thin, simple, toothed leaves.
- Two-lipped flowers, scarlet with yellowish hairy throat; many small seeds in a cylindrical capsule; blooms April to October.

FIGWORT FAMILY *Scrophulariaceae*

Monkey Flower

Mimulus guttatus

- Exceedingly common in wet places below 10,000 feet; many plant communities in most of California.
- Usually perennial, rooting at the nodes of creeping rootstock or stolon; 2 to 40 inches tall.
- Oval, rounded, toothed leaves.
- 2-lipped flowers, yellow with red spots and a hairy throat, variable; fruit many small seeds in a cylindrical capsule; blooms March to August.

Muskflower *M. moschatus*, growing in wet places below 7500 feet, is a hairy perennial with slender creeping rootstocks, leaves with a slimy or slippery feeling, and 2-lipped yellow flowers with hairy throat; blooms June to August.

PURSLANE FAMILY *Portulacaceae*

Siberian Candyflower

Montia sibirica

- Moist shaded places, below 5000 feet; Redwood, Douglas-Fir, Red Fir, and Mixed Evergreen forests, Northern Coastal Scrub; Santa Cruz to Del Norte and Siskiyou counties.
- Rather succulent perennial with few to several weakish stems, 6 to 18 inches high.
- Several basal, ovate leaves, 2 smaller stem leaves.
- Flower petals white with pink lines, capsular fruit with 1-3 black shiny seeds; blooms March to August.

GRASS FAMILY *Poaceae*

Deergrass

Muhlenbergia rigens

- Dry or damp places below 7000 feet; Valley Grassland, Chaparral, Yellow Pine Forest; foothills of Sierra Nevada from San Diego County north.
- Tall, slender, perennial bunchgrass, 28 to 60 inches tall.
- Leaf blades narrow, somewhat roughened, rolling inward.
- Inflorescence slender of 1-flowered spikelets without bristles; blooms June to September.

WAX-MYRTLE FAMILY *Myricaceae*

Wax-Myrtle

Myrica californica

- Canyons and moist slopes below 500 feet, Redwood Forest, Coastal Strand, Northern Coastal Scrub, Santa Monica County north.
- Evergreen shrubs, 6 to 12 feet tall, or small trees to 30 feet.
- Dark green, glossy, persistent leaves.
- Male and female flowers in separate short scaly catkins on the same plant, maturing into a purplish-brown nut with a fleshy outer-coat covered with whitish wax; blooms March-April.

Sierra Sweet-bay *Myrica hartwegii* is a deciduous shrub, 3-6 feet tall,growing along stream banks in the Yellow Pine Forest of the Sierra Nevada with unisexual flowers on separate plants; blooms May to June.

CARROT FAMILY *Apiaceae*

Water Parsley

Oenanthe sarmentosa

- Marshes and sluggish water below 3500 feet; many plant communities west of the Sierra Nevada.
- Weak-stemmed perennial; 20 to 60 inches tall.
- Leaves pinnately compound with coarsely-toothed leaflets.
- Many, small, white flowers in an umbrella-like inflorescence, developing into ribbed purplish-green fruit; blooms June to October.

PEA FAMILY *Fabaceae*

Desert-Ironwood

Olneya tesota

- Desert washes below 2000 feet; Creosote Bush Scrub in the Colorado Desert.
- Broad-crowned, greyish, spinose tree with thin scaly bark and pairs of spines below the leaves, 15 to 24 feet tall.
- Leaves with 8-24 small leaflets.
- Pale rose-purple pea-shaped flowers in a long inflorescence (raceme), fruit a thick pod with constrictions, seeds black; blooms April to May.

OXALIS FAMILY *Oxalidaceae*

Redwood-Sorrel

Oxalis oregana

- Shady creeksides and redwood banks; Redwood and Douglas-fir forests, Monterey County north to Oregon.
- Annual or perennial, ground cover, with branching rootstock growing in shade; generally under 12 inches tall.
- 3 or more leaflets folding at night.
- White or pinkish petals often veined purple, fruit a small capsule; blooms April to September.

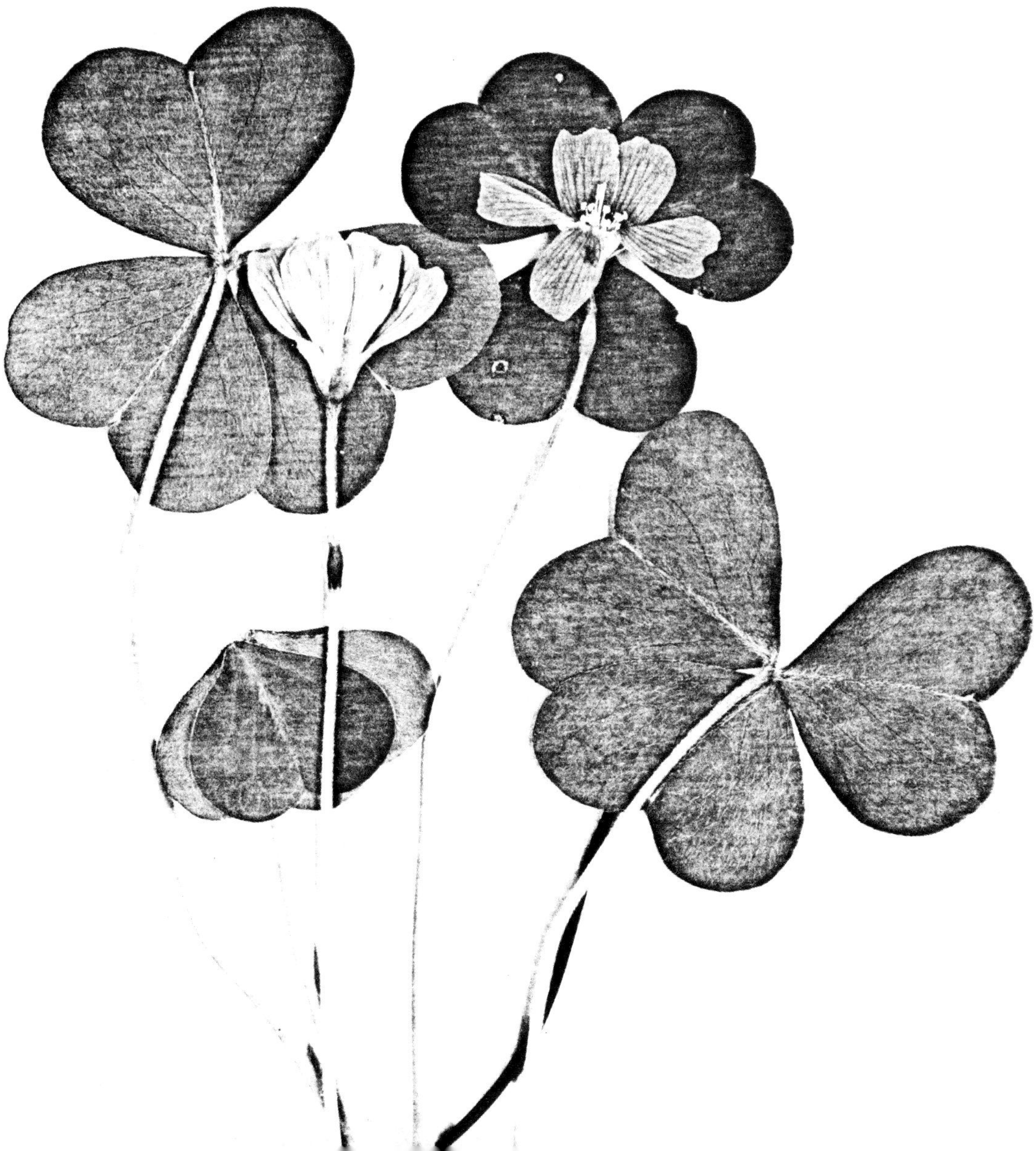

SUNFLOWER FAMILY *Asteraceae*

Western Coltsfoot

Petasites palmatus

- Deep shade in wooded places and along streams below 1000 feet; Mixed Evergreen and Redwood forests; Coast Ranges from Santa Lucia mountains to Del Norte and Siskiyou counties.
- Perennial herb with creeping rootstock, 4 to 20 inches tall.
- Large, roundish, basal leaves, densely hairy beneath.
- Whitish flowers in dense terminal clusters, subtended by large leaflike bracts, on stems up to almost 2 feet tall, arising in early spring, basal leaves emerge later; fruit an akene with thread-like bristles; blooms March to April.

SAXIFRAGE FAMILY *Saxifragaceae*

Mock-Orange

Philadelphus lewisii subsp. *californicus*

- Rocky slopes and canyons, 1000 to 4500 feet; Foothill Woodland and Yellow Pine Forest in Sierra Nevada from Tulare County north to Humboldt, Siskiyou and Trinity counties.
- Loosely-branched, deciduous shrub, 3 to 9 feet tall.
- Opposite, entire, toothed leaves.
- Showy white flowers with many stamen, fruit a many seeded capsule; blooms May to July.

Named for Ptolemy, *Philadelphus* of ancient Egypt.

MISTLETOE FAMILY *Loranthaceae*

Mesquite Mistletoe 

Phoradendron californicum

- A parasitic woody plant found mostly on *Prosopis*, occasionally on *Acacia*, *Cercidium*, *Tamarix*, and *Larrea* in the deserts of Southern California.
- Leaves reduced to scales.
- Flowers sunk into stem, develop into round waxy berries that are actively spread from host to host by birds.

P. flavescens is common on cottonwood, ash, sycamore, walnut and willows, especially in the Sacramento and northern San Joaquin valleys.

GRASS FAMILY *Poaceae*

Giant Reed

Phragmites australis

- Sunny, sandy, sub-irrigated sites from the Delta to the desert where water is available; in many plant communities below 5000 feet.
- Cosmopolitan perennial reeds growing from long creeping rootstocks and forming cane-like thickets, 6 to 24 feet tall.
- Leaf blades 0.5 to 2 inches wide.
- Large terminal floral plume.

ROSE FAMILY *Rosaceae*

Ninebark

Physocarpus capitatus

- Moist banks and north slopes below 4500 feet; many plant communities from Redwood to Red Fir forests; Sierra Nevada from Tulare County north, Coast Ranges from Santa Barbara County north.
- Erect or spreading, deciduous shrub with shredding bark, 3 to 7 feet tall.
- Roundish, palmately lobed, and toothed leaves.
- Dense clusters of small white flowers which mature into inflated, dry, 1-chambered fruit; blooms May to July.

SYCAMORE FAMILY *Platanaceae*

Sycamore

Platanus racemosa

- Common along perennial and intermittant streams and water courses below 4000 feet, particularly where substrate is cobbly or bouldery; many plant communities of the Great Valley and bordering mountains.
- Large deciduous tree with smooth grey peeling bark, 30 to 75 feet tall.
- 5-lobed palmate leaves, both sides hairy when young.
- Minute flowers clustered into 1/3 inch spherical balls on long stalks, sexes separate but on same tree, fruit in dense globose hairy head of 1-seeded dry akenes; blooms February to April.

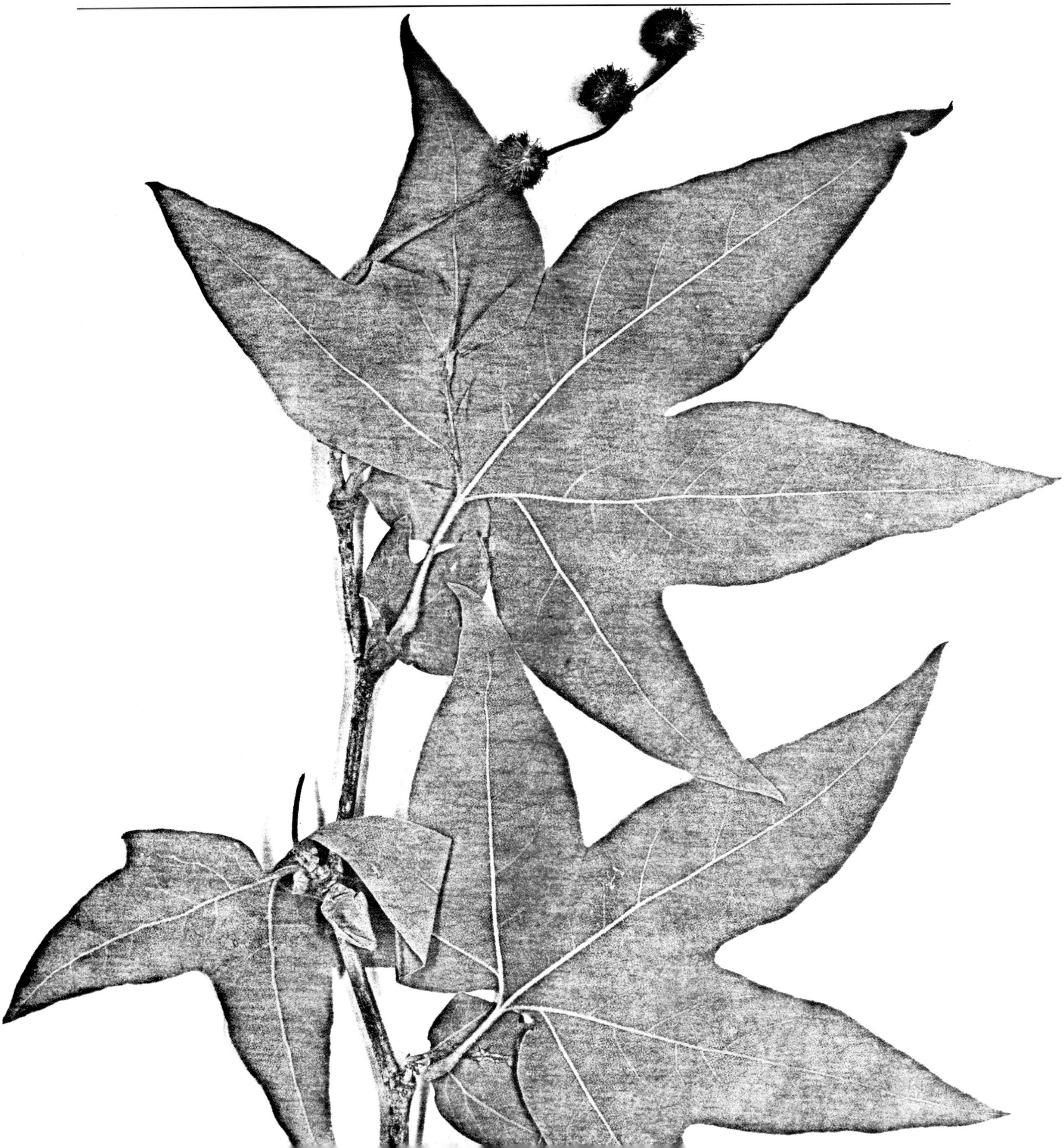

SUNFLOWER FAMILY *Asteraceae*

Arrowweed

Pluchea sericea

- Frequent in sandy subirrigated places, washes, and river bottoms; Coastal Sage and Creosote Bush Scrub, Southern California north to Santa Barbara County.
- Tall leafy willow-like shrubs, 3 to 12 feet tall.
- Alternate, silvery-silky, lance-shaped leaves.
- Rayless purplish flowers in numerous terminal heads, akene fruits bear capillary bristles; blooms March to July.

FERN FAMILY *Polypodiaceae*

Polypody

Polypodium californicum

- Common on rocky ledges and moist banks below 4000 feet; Chaparral, Coastal Sage Scrub, Foothill Woodland, Mixed Evergreen Forest; west of the Sierra Nevada through foothills and Coast Ranges.
- Epiphytic, without soil roots, but with creeping rhizomes.
- Fronds not evergreen, blades deeply divided nearly to center.
- Fruiting dots (sori) on underside of fronds.

FERN FAMILY *Aspidiaceae*

Sword Fern

Polystichum munitum

- Common in damp woods, mostly below 2500 feet; Redwood, Northern Coniferous, Douglas-fir forests near the coast from Monterey to Del Norte counties.
- Coarse evergreen fern with many fronds from woody scaly rhizomes; 24 to 56 inches tall.
- Fronds divided once into dense and evenly-laced pinnae, individual pinna toothed and sword shaped with a larger lobe at the base like a sword hilt.
- Sori arranged in 2 rows near margin of pinnae each covered by a round indusium.

WILLOW FAMILY *Salicaceae*

Fremont Cottonwood

Populus fremontii

- Moist places below 5600 feet, several plant communities west of Sierra Nevada, in Southern California and Mojave Desert, Central Valley and bordering hills to Lake County.
- Tall open-crowned tree with whitish roughly-cracked bark, 40 to 90 feet tall, separate male and female trees.
- Leaves coarsely-toothed and triangular, same color on both sides, leaf stem (petiole) flattened on sides so leaves "quake" in the wind.
- Inconspicuous flowers in male or female catkins, 1.5 to 2 inches, fruit a dry capsule with many minute seeds bearing tufts of hair; blooms March to April.

Quaking aspen *P. tremuloides* grows along stream borders and dampish places, 6000 to 10,000 feet in the San Bernardino, Sierra Nevada and Trinity mountains; its leaves are ovate to round, toothed with pointed tip, and leaf stems are flattened laterally so leaves "quake" in the wind. Fremont cottonwood was named for Major John Charles Fremont, who with Kit Carson, discovered it near Pyramid Lake in 1844. They made good use of its riparian affinity for locating water.

WILLOW FAMILY *Salicaceae*

Black Cottonwood

Populus trichocarpa

- Common along streams below 9000 feet; many plant communities throughout California west of the Sierra Nevada.
- Tall, open-crowned, deciduous tree with greyish bark becoming furrowed in age, separate male and female trees; 120 to 180 feet tall.
- Finely toothed ovate leaves on long stems, dark green above and pale beneath.
- Inconspicuous flowers in male or female catkins, 1.5 to 3 inches, on separate trees, fruit a dry capsule with many minute seeds bearing a tuft of hairs; blooms February through April.

Black cottonwood, our tallest cottonwood and common in much of California outside the Great Valley, is more shallow rooted and grows in shallower soils than *P. fremontii*.

PEA FAMILY *Fabaceae*

Honey Mesquite

Prosopis glandulosa var. *torreyana*

- Common in washes, low places, and sandy sites below 3000 feet; Creosote Bush Scrub, edge of Alkali Sink, Colorado and Mojave Desert, upper San Joaquin Valley, interior San Bernardino County west to San Diego region.
- Deciduous low tree or large shrub with several trunks and crooked arched branches, 5 to 20 feet tall.
- Leaves twice-divided into bright-green paired leaflets.
- Flowers small, greenish to yellow in slender spikes, pods flat, several in a cluster, constricted between seeds; blooms April through June.

PEA FAMILY *Fabaceae*

Screwbean, Mesquite

Prosopis pubescens

- Washes and canyons below 2500 feet; Mojave and Colorado desert.
- Deciduous shrub or small tree to 30 feet tall with pale slender twigs and stout spines.
- Greyish downy covered leaves twice divided into numerous, oblong, paired leaflets.
- Small yellow flowers in long spikes, seed pods tightly coiled; blooms May to July.

Where honey and screwbean mesquite occur together, screwbean prefers sites with somewhat deeper soils and tends to occur on outer floodplain margins. Pods of both species are much eaten by cattle.

PEA FAMILY *Leguminosae*

Psoralea

Psoralea macrostachya

- Moist places below 5000 feet; many plant communities in most of California west of the Sierra Nevada.
- Heavy-scented perennial herb, 2 to 9 feet tall.
- Leaves with 3 rounded leaflets, obscurely gland-dotted.
- Long spikes of small, purple, pea-shaped flowers; fruit an egg-shaped pod; blooms May to August.

PEA FAMILY *Fabaceae*

Smoke Tree, Smokebush

Psorothamnus spinosa

- Common in sandy washes below 1500 feet; southern Mojave and Colorado deserts.
- Intricately-branched, smokey-grey, spiny shrub or small tree, 3 to 18 feet tall, dotted with glands thoughout.
- Leaves small, simple, and drought deciduous, mostly leafless.
- Tiny blue-purple flower clusters, fruit small ovoid pods covered with amber-colored glands; blooms June to July.

Though this plant is leafless for much of the year, considerable photosynthesis is carried on in the chlorophyll-containing stems.

BEECH FAMILY *Fagaceae*

Valley Oak

Quercus lobata

- On rich loam in floodplains, valleys, and most slopes below 2000 feet; Foothill Woodlands, Great Valley and its borders, Inner and Middle Coast Ranges to Los Angeles County.
- Stately, graceful, deciduous tree with open head, 35 to 100 feet tall with trunk to 12 feet thick, bark thick in cuboidal plates.
- Leaves with 3-5 pair of rounded lobes, paler beneath.
- Separate male catkins, female flowers develop into long conical acorns with round cups; blooms March to April.

Purported to be the largest American oak, the valley oak is not restricted to riparian habitat everywhere but is considered endangered by the loss of riparian habitat and by extensive cattle grazing in California.

BUCKTHORN FAMILY *Rhamnaceae*

Cascara Sagrada

Rhamnus purshiana

- Moist places below 5000 feet; Redwood, Mixed Evergreen, Yellow Pine forests, Sierra Nevada foothills from Placer County north, Outer Coast Ranges from Lake and Mendocino counties north.
- Large deciduous shrub or small tree, 15 to 36 feet tall.
- Leaves alternate, tufted at ends of branches, broad ovoid in shape.
- Small clusters of tiny greenish flowers in leaf axils, berries black; blooms May to July.

Bark has laxative properties.

HEATH FAMILY *Ericaeae*

Western Azalea

Rhododendron occidentale

- Stream banks and moist places below 7500 feet; Yellow Pine, Red Fir, Mixed Evergreen and Redwood forests; Cuyamaca to San Jacinto mountains, Sierra Nevada from Kern County north, Coast Ranges from Santa Cruz north.
- Loosely-branched deciduous shrub, 3 to 15 feet tall.
- Thin, light-green, long leaves.
- Flowers, fragrant, large and showy, irregularly-lobed, white or with pinkish tinge, prominant anthers; fruit many scale-like seeds in a dry capsule; blooms April to August.

SAXIFRAGE FAMILY *Saxifragaceae*

Golden Currant

Ribes aureum

- Moist banks, 2500 to 7800 feet, Sagebrush Scrub to Lodgepole Forest, Fresno and Inyo Counties, then east of Sierra Nevada to Modoc County, and occasional in river-bottom lands in theSan Joaquin Valley.
- Unarmed erect shrub, 3 to 6 feet tall.
- Rounded, 3-lobed leaves with toothed edges.
- Small clusters of flowers with yellow petals fading to orange and a spicy fragrance, developing into orange, red, or black berries; blooms April to May.

The golden currant has a wide range over the west; the type specimen was collected by Meriwether Lewis in the Rocky Mountains during the Lewis and Clark expedition of 1804-1806.

SAXIFRAGE FAMILY *Saxifragaceae*

Red Flowering Current

Ribes sanguineum
var. *glutinosum*

- Open places among shrubs or trees below 2000 feet; Chaparral, Mixed Evergreen and Closed-cone Pine forests, Foothill Woodland in Coast Ranges from Santa Barbara to Del Norte counties.
- Erect, deciduous shrub, 3 to 9 feet tall with slender stems.
- Thick, roundish, wrinkled leaves with stalked glands.
- Deep to pale pink flowers in short clusters (racemes) developing into blackish spineless berries; blooms March to April.

Red flowering current *R. sanguineum* occurring in moist shaded places, 2000 to 6000 feet, in Yellow Pine and Red Fir forests, has deep rose flowers in small clusters developing into black berries with a bloom.

SAXIFRAGE FAMILY *Saxifragaceae*

Fushia-Flowered Gooseberry

Ribes speciosum

- Common in shaded canyons below 1500 feet; Coastal Sage Scrub and Chaparral near the coast from San Diego to Santa Clara counties.
- Evergreen shrub, long, spreading, bristly branches with nodal spines, 3 to 6 feet tall.
- Leaves roundish, somewhat 3-lobed, shining dark-green above, paler beneath.
- Flowers red, bellshaped, ovary glandular-hairy developing into glandular-bristly berries; blooms January to May

PEA FAMILY *Fabaceae*

Locust

Robinia pseudo-acacia

- A widely planted street tree, now naturalized in many floodplains, particularly in Northern California.
- Deciduous tree with rough bark, spines at leaf bases, up to 70 feet tall.
- Pinnately compound leaves with 9-19 thin ovate leaflets.
- Showy, white, fragrant, drooping clusters of pea-like flowers, developing into long flat pods; blooms May to June.

MUSTARD FAMILY *Brassicaceae*

Water-Cress

Rorippa nasturtium-aquatica

- Common in quiet waters and wet banks below 8000 feet; many plant communities in California.
- Aquatic perennial with prostrate or ascending stems 4 to 24 inches long, rooting freely.
- Leaves pinnate with 3 to 11 oval leaflets.
- Tiny white clusters of flowers, fruit a many-seeded linear capsule opening by 2 valves; blooms March to November.

Water-cress was naturalized from Europe; the same species is grown commercially.

ROSE FAMILY *Rosaceae*

California Wild Rose

Rosa californica

- Fairly moist places in canyons and near streams below 6000 feet; many plant communities in California west of the Sierra Nevada except the deserts.
- Erect, branched shrub, armed with stout, flattened, usually-recurved prickles; 3 to 9 feet tall.
- Leaves composed of 5-7 oval slightly-toothed leaflets.
- Rose-pink flowers in clusters, fruit a round red hip containing hairy 1-seeded akenes and high in Vitamin C; blooms May to August.

ROSE FAMILY *Rosaceae*

Thimbleberry

Rubus parviflorus

- Open woods and in canyons below 8000 feet; Red Fir, Yellow Pine, and Redwood forests from San Diego County north, Sierra Nevada and Coast Ranges.
- Deciduous shrub 3 to 6 feet high, without prickles, with shreddy bark in age.
- Leaves 5-lobed, soft, and unequally toothed.
- Few, large, white flowers in terminal clusters, fruit an edible aggregation of numerous small red stone fruits separating at maturity from the raised receptacle; blooms March to August.

ROSE FAMILY *Rosaceae*

Salmon Berry

Rubus spectabilis

- Moist places in and about woods below 1000 feet; Redwood and Mixed Evergreen forests; Mendocino to Del Norte Counties.
- Branching, glandless, leafy shrub, 6 to 12 feet tall, primocanes sometimes with small straight prickles, older canes with shredding bark.
- Deciduous leaves with 3 doubly-toothed leaflets.
- Good-sized red-purple flowers developing into edible, red to salmon or yellow, round to cone-shaped aggregations of small stone fruits; blooms March to June.

ROSE FAMILY *Rosaceae*

California Blackberry

Rubus vitifolius,
including *R. ursinus.*

- Woods, canyons, waste and damp places below 4000 feet; near the coast from Mendocino to San Luis Obispo counties, most of California west of the Sierra Nevada.
- Mound-building or trailing prickly shrub with tip-rooting running stems; flowering shoots to 20 feet.
- Leaves 3-foliate, ovate, doubly-toothed but variable.
- Flower petals white with many stamen; fruit an elongate aggregate of black stone fruits, highly edible; blooms March to June.

Loganberry, youngberry, and boysenberry have all been developed from *R. vitifolius*. Himalaya berry *R. procerus* is a robust, sprawling, European escape, becoming naturalized in Northern California in disturbed waste places. It has 5 leaflets in primocane leaves, 3-5 leaflets in leaves of older floricanes.

WILLOW FAMILY *Salicaceae*

Geyer's Silver Willow

Salix geyeriana var. *argentea*

- Wet places 5000 to 10,500 feet; Montane Coniferous Forest; Sierra Nevada from Tulare County north to Modoc County.
- Shrub to 3 feet tall, with blackish twigs.
- Leaves narrow, silvery, hairy on both sides, pointed at both ends, stipules lacking.
- Catkins emerge with leaves on separate shrubs; blooms in May.

The medicinal value of willow bark and leaves is due to the presence of a bitter, white, crystalline glucoside, salicin, from which is derived salicylic acid, which in turn yields aspirin.

WILLOW FAMILY *Salicaceae*

Goodding's Willow

Salix gooddingii var. *variabilis*

- Abundant in wet places, Valley Grasslands, Foothill Woodlands, Coastal Sage Scrub; Great Valley from Shasta County south.
- Tree with rough dark bark and yellowish twigs, 18 to 30 feet tall.
- Leaves narrow, finely glandular and toothed, greyish-green on both sides, glandular stipules.
- Male and female catkins 11/2 to 3 inches long, on separate trees; blooms March to April.

WILLOW FAMILY *Salicaceae*

Sandbar Willow

Salix hindsiana

- Common locally along ditches and on sandbars below 3000 feet; many plant communities throughout California west of the Sierra Nevada; blooms March to May.
- Erect shrub or small tree with grey furrowed bark, 6 to 21 feet tall.
- Narrow linear leaves generally covered by grey silky hair.
- Catkins on long leafy stems, appear after leaves; blooms March to May.

Narrow-leaf willow *S. exigua* replaces *S. hindsiana* in wet places below 8000 feet, Sagebrush and Creosote Bush Scrub, deserts and east of Sierra Nevada; erect shrub or small tree, 6 to 12 feet; blooms March to May.

WILLOW FAMILY *Salicaceae*

Red Willow

Salix laevigata

- Along streams below 5000 feet; many plant communities west of the Sierra Nevada.
- Tree with rough bark and reddish-brown twigs, 15 to 45 feet tall.
- Leaves oblong-lanceolate shaped, finely toothed, light green above, paler and cloudy beneath, small glandular-toothed stipules.
- Male and female catkins on separate trees, 1 to 4 inches; blooms March to May.

WILLOW FAMILY *Salicaceae*

Yellow Tree Willow

Salix lasiandra

- Stream banks below 8000 feet; many plant communities through most of California except deserts.
- Treelike growth with yellow to reddish twigs, 20 to 50 feet tall.
- Leaves dark green, shining above with characteristic wart-like glands on leaf stems.
- Separate male and female trees, female catkins longer; blooms March through May

WILLOW FAMILY *Salicaceae*

Arroyo Willow

Salix lasiolepis

- Common on stream banks and beds below 7000 feet, many plant communities throughout California west of the Sierra Nevada.
- Erect shrub or small tree, 6 to 30 feet tall.
- Long narrow leaves on yellowish twigs.
- Male and female flowers on separate trees, inconspicuous flowers in catkins.

S. hookeriana, ocean beaches above stream mouths and lagoons, below 500 feet, Mendocino to Del Norte counties; *S. melanopsis*, stream banks below 8000 feet in the Sierra Nevada and the Coast Ranges from Lake and Sonoma counties north; *S. piperi*, wet sandy places at mouths of streams below 500 feet, Mendocino County north; *S. tracyi*, sand and gravel bars of streams below 500 feet, Humboldt and Del Norte counties; *S. parksiana*, gravel bars and along streams below 1000 feet, Humboldt and Del Norte counties.

WILLOW FAMILY *Salicaceae*

Yellow Willow

Salix lutea

- Wet places along streams, 5000 to 9500 feet; Red Fir and Lodgepole forests, Sierra Nevada from Nevada to Modoc counties.
- Clustered shrub 6 to 15 feet tall, with yellowish or brownish twigs.
- Leaves narrow lance-shaped, rounded at base, some may be finely toothed, stipules oval to moon-shaped.
- Male and female catkins appear before and with leaves, 3/4 to 1-1/4 inces long on separate plants; blooms May to June.

Brewers' Willow *S. breweri*, growing along gravel beds and ravines below 4500 feet, Foothill Woodlands, Inner Coast Ranges, San Benito to Lake and Colusa counties, is a shrub 3 feet tall; blooms March to April.

HONEYSUCKLE FAMILY *Caprifoliaceae*

Red Elderberry

Sambucus callicarpa

- Damp woods and floodplains at low elevations; Northern Coastal Scrub, Redwood and Douglas-fir forests, along the coast from San Mateo County north.
- Shrubs or small trees, 6 to 18 feet tall.
- Opposite compound leaves with 5-7 toothed leaflets, 2 leaves at each node.
- Whitish flowers in dome-shaped clusters that form bright scarlet, berry-like, somewhat bitter-tasting fruit, sometimes toxic; blooms March to July.

HONEYSUCKLE FAMILY *Caprifoliaceae*

Blue Elderberry

Sambucus cerulea

- Open places up to 10,000; largely Montane Coniferous Forests from San Diego County north through the Sierra Nevada and the Coast Ranges.
- Large shrub or small tree, 6 to 24 feet tall.
- Leaves opposite, compound, sharply toothed with 5 to 9 leaflets, 2 leaves at each node.
- Whitish flowers in large flat-topped clusters; berries blackish with a whitish bloom, thus appearing blue; blooms March to September.

This is the host plant for the Valley Elderberry Longhorn Beetle, designated endangered under the Endangered Species Act.

SEDGE FAMILY *Cyperaceae*

Bulrush, Scirpus

Scirpus microcarpus

- Occasional along streams and about springs in mountains up to 9000 feet; many plant communities west of Sierra Nevada from San Diego County north.
- Perennial with stout rhizomes, stems erect and roundish, 3 to 6 feet tall.
- Flat broad leaves growing along stem.
- A loose spreading cluster of petaless flowers surrounded by a series of green to brown scales, fruit minute lensshaped akenes; blooms May to August.

The genus *Scirpus* is large with over 200 species distributed worldwide and 17 species distributed throughout California.

FIGWORT FAMILY *Scrophulariaceae*

California Bee-Plant

Scrophularia californica

- Common in damp places, especially brushy thickets; Closed-cone Pine and Redwood forests; near coast from Santa Monica Mountains north to Oregon.
- Coarse 4-angle-stemmed perennial, 40 to 72 inches tall.
- Triangular-ovate, opposite, and toothed leaves.
- Red-brown to maroon, smallish tubular 2-lipped flowers, many seeds in a small dry capsule; blooms February to July.

SUNFLOWER FAMILY *Asteraceae*

German Ivy

Senecio mikanioides

- Naturalized aggressive weedy species in canyons and gullies along coast of Southern and Central California.
- Leafy perennial with slender twining stems that tend to blanket other vegetation.
- Roundish, thin, ivy-shaped leaves.
- Flowers in condensed clusters in leaf axils toward summit of stems, fruit an akene with many soft white bristles; blooms in March.

Native of South Africa

LILY FAMILY *Liliaceae*

False Solomon's-Seal

Smilacina stellata var. *sessilifolia*

- Moist slopes and streambanks, Chaparral, Redwood and Mixed Evergreen forests; Coast Ranges, Santa Barbara County north.
- Perennial herb with creeping rootstock, 12 to 24 inches high.
- Leaves alternate, little leaf-stalk, 3 veins generally more prominent.
- Clusters of small white flowers in an elongate, terminal, branched inflorescence (raceme), fruit a red-purple berry aging to black; blooms April to June.

Smilacina racemosa var. *amplexicaulis* grows in shaded woods below 6000 feet, mostly in Mixed Evergreen and Montane Coniferous forests through the Coast Ranges and the Sierra Nevada. It has clusters of small white flowers on an elongate, terminal twice branched inflorescence (panicle), fruit a red berry, sometimes with purple spots.

LILY FAMILY *Liliaceae*

Greenbriar

Smilax californica

- Thickets and stream banks below 5000 feet; Yellow Pine and Mixed Evergreen forests; Trinity to Siskiyou and Butte counties.
- Woody, prickly-stemmed, straggling vine; 3 to 9 feet.
- Ovate heart-shaped leaves.
- Unisexual greenish or yellowish small flowers growing in umbrella-like clusters from axils of leaves of separate plants, and developing into small black berries; blooms May to June.

PARSLEY FAMILY *Apiaceae*

Ranger's Button

Sphenosciadium capitellatum

- Swampy places 3000 to 10,400 feet, Yellow Pine to Subalpine forests; San Jacinto and San Bernardino Mountains, Mt. Pinos, Tehachapi Mountains, Sierra Nevada, White and Sweetwater mountains.
- Tall perennial herb, 20 to 50 inches tall; leaves divided into large leaflets with sheath like leaf-stalks.
- White or purplish flower clusters in umbrella-like round heads, small, dry, 2-parted fruits, ribbed at base and winged above; blooms July to August.

ROSE FAMILY *Rosaceae*

Spiraea

Spiraea douglasii

- Low damp places below 6000 feet, Redwood to Red Fir forests; from Humboldt, Trinity, Butte and Plumas Counties north.
- Erect shrub, 3 to 6 feet tall.
- Simple toothed leaves.
- Small rose-colored flowers congested on terminal, elongate, branched inflorescence, small, dry, 1-chambered fruits; blooms from June to September.

S. douglasii is replaced at higher elevations by *S. densiflora*.

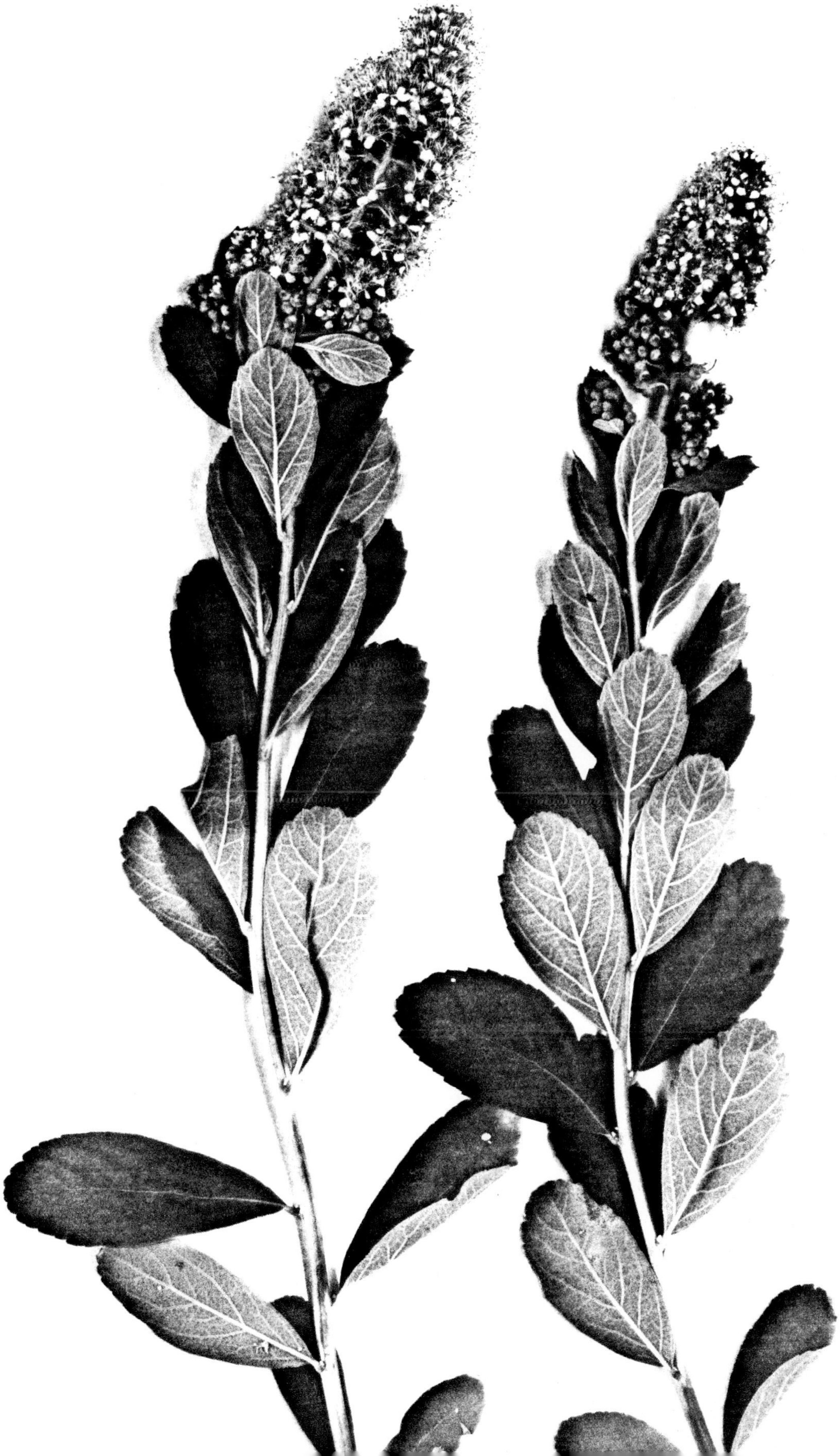

MINT FAMILY *Lamiaceae*

Hedge-Nettle

Stachys chamissonis

- Moist places at low elevations near the coast; Redwood and Mixed Evergreen forests, Northern Coastal Scrub; San Luis Obispo, San Mateo to Humboldt Counties.
- Stout square-stemmed perennial, 24 to 40 inches tall.
- Narrow ovate leaves coarsely toothed and covered with bristly hairs.
- Rose-purple tubular flowers in interrupted whorls on spike, fruit of 4 1-seeded nutlets in a persistent calyx; blooms June to October.

HONEYSUCKLE FAMILY *Caprifoliaceae*

Snowberry

Symphoricarpos rivularis

- Banks and flats in canyons near streams below 4000 feet; many plant communities in the Coast Ranges from Monterey north, and the Northern Sierra Nevada.
- Erect branched shrub, 1 to 6 feet tall.
- Leaves oval and toothed, opposite and deciduous.
- Rose-pink to white, short, tubular flowers, fruit a small white, berry-like, stone fruit with 2 nutlets; blooms May to July.

TAMARISK FAMILY *Tamaricaceae*

Tamarisk

Tamarix sp.

- Water courses and low places below 4000 feet, coastal and desert drainages particulary common in Southern California; loosely branched shrub or small tree, several difficult to distinguish species, 3 to 9 feet tall.
- Thickish scale-like leaves.
- Abundant small white or pinkish flowers in slender clusters, many tufted seeds in small dry capsules; blooms April through August (some year-round).

An escaped exotic that displaces natives, Tamarix has invaded watercourses of the west with remarkable speed (from unknown to common in 20 years) and is considered a major pest species.

SAXIFRAGE FAMILY *Saxifragaceae*

Fringe-Cups

Tellima grandiflora

- Moist woods and rocky places below 5000 feet; Redwood, Mixed Evergreen, and Yellow Pine forests; Coast Ranges from San Luis Obispo County north, and Sierra Nevada from Placer County north.
- Perennial with numerous basal leaves, 16 to 32 inches tall.
- Roundish, shallowly-lobed, toothed leaves, smaller on stout flowering stem.
- Flowers on long, terminal, branched, many-flowered inflorescences, petals with fringed edge, opening whitish fading to red, fruit a dry capsule opening between the 2 narrow beaks; blooms April to June.

BUTTERCUP FAMILY *Ranunculaceae*

Meadow-Rue

Thalictrum polycarpum

- Moist soil near streams and meadows, often in shade and thickets, 4000 to 10,000 feet; Montane Coniferous Forests, throughout the Sierra Nevada, near the coast from Central California north.
- Slender perennial, 24 to 60 inches tall.
- Delicate divided leaves, minutely glandular and hairy, foliage has unpleasant odor when crushed.
- Male and female flowers on different plants, delicate greenish with no petals; blooms May to August.

SAXIFRAGE FAMILY *Saxifragaceae*

Tolmiea

Tolmiea menziesii

- Moist cool places below 6000 feet; Red Fir, Mixed Evergreen, Redwood, Douglas-Fir forests; Glenn County north.
- Perennial herb with scaly rootstock and basal leaves.
- Leaves somewhat-lobed, toothed, and hairy.
- Long floral infloresence (raceme) with many greenish tubular flowers, some with purple-brown petals, fruit a 1-celled double-beaked capsule with many seeds; blooms May to June.

SUMAC FAMILY *Anacardiaceae*

Poison Oak

Toxicodendron diversilobum

- Common in low places and thickets and wooded slopes below 5000 feet, many plant communities west of the Sierra Nevada.
- Erect shrub, 3 to 6 feet tall; or a vine climbing high into the canopy of many species of riparian trees.
- Three-parted lobed leaflets.
- Inconspicuous whitish flowers, stone-fruit with 1 seed; blooms April-May.

Causes a painful dermititis to humans; fruits are a very important food source for many wildlife species.

CATTAIL FAMILY *Typhaceae*

Cattail

Typha domingensis

- Widely distributed in subsaline or freshwater marshes below 5000 feet throughout California.
- Tall perennial with creeping rotstock; 6 to 9 feet tall.
- 6 to 9 long straplike leaves, somewhat convex on underside.
- Flowers crowded on terminal spike with male flowers above, cinnamon-brown female flowers below, small dry fruits with copius basal hairs; blooms June and July.

Narrow-leaf cattail *T. angustifolia* and soft-flag *T. latifolia* are also common species in freshwater sites in California.

LAUREL FAMILY *Lauraceae*

California-Bay

Umbellularia californica

- Common in canyons and valleys below 5,000 feet; Chaparral, Foothill Woodlands, lower Yellow Pine, Mixed Evergreen, Redwood forests in the Coast Ranges and the Sierra Nevada, San Diego County north; more restricted to riparian corridors in the southern part of its range.
- Tree with broad crown to 90 feet, or erect shrub in drier places.
- Long, oblong, aromatic leaves.
- Clusters of small yellowish-green flowers, lacking petals and bearing 6 orange glands, maturing into greenish stone-fruit which ripens purple; blooms December to May.

California bay is in the same family as avocado, cinnamon, and camphor.

NETTLE FAMILY *Urticaceae*

California Nettle

Urtica californica

- Wet brushy thickets along the coast; Central California north.
- Perennial herbs with stinging hairs on leaves, stems, and flowers.
- Leaves opposite, simple, and toothed, densely hairy and greyish beneath.
- Small greenish male and female flowers in separate clusters, hanging from axils of upper leaves, fruit a small, dry, 1-seeded akene; blooms April to June.

U. holosericea, occuring in low damp places below 9000 feet, is common throughout California west of the Sierra Nevada Range; stems are whitish with short hairs, leaves twice as long as wide; blooms June to September.

HEATH FAMILY *Ericaceae*

California Huckleberry

Vaccinium ovatum

- Common on dry slopes and canyons below 2500 feet; Redwood, Closed-cone Pine, and Mixed Evergreen forests near coast from Santa Barbara County north.
- Stout, erect, much branched shrub, 3 to 7 feet tall.
- Small, leathery, ovate leaves, shining above, pale beneath.
- Small, white to pink, bell-shaped flowers in clusters developing into sweet edible black berries; blooms March to May.

VERBENA FAMILY *Verbenaceae*

Vervain

Verbena hastata

- Moist waste places along Sacramento and San Joaquin Rivers.
- Erect perennial, 2 to 5 feet tall.
- Leaves deeply-indented and toothed.
- Small blue, pink or white tubular flowers crowded on short spikes, fruit of 4 nutlets; blooms June to September.

FIGWORT FAMILY *Scrophulariaceae*

American Brooklime

Veronica americana

- Common in along streams and near springs below 10,500 feet; many plant communities west of the Sierra Nevada.
- Somewhat succulent perennial, creeping at the base from rhizomes, with erect or decumbent stems, 4 to 40 inches tall.
- Leaves lance-shaped, pointed, and toothed.
- Violet-blue flowers with 2 stamen, in leaf axils; fruit a many-seeded capsule; blooms May to August.

GRAPE FAMILY *Vitaceae*

Wild Grape

Vitis californica

- Stream banks and canyons, below 4000 feet; Mixed Evergreen Forest, Northern Oak and Foothill woodlands; Coast Ranges from San Luis Obispo County north.
- Woody vine climbing by tendrils opposite the leaves.
- Leaves alternate, roundish, palmately lobed or toothed, hairy particularly beneath.
- Flowers small, greenish-yellow, and fragrant, in clusters developing into purple fruit; blooms May to June.

PALM FAMILY *Palmae*

California Fan Palm

Washingtonia filifera

- In groves in moist alkaline spots around seeps, springs and streams below 3500 feet; Creosote Bush Scrub; west and north edge of Colorado Desert, Turtle Mountains, region of Twenty-nine Palms of Mojave Desert.
- Tall columnar tree commonly clothed with dense thatch of dead drooping leaves, 30 to 45 feet tall.
- Leaves large, heavy, with 40 to 60 folds, segmented to the middle, copiously fibrous.
- Flowers small in clusters surrounded by large sheathing bract (spathe), fruit a drupe, hard, fleshy, and berry-like; blooms in June.

Named in honor of George Washington.

FERN FAMILY *Blechnaceae*

Chain Fern

Woodwardia fimbriata

- Springy, boggy places in canyons below 5000 feet; many plant communities west of Sierra Nevada, occasional along desert edge.
- Stout, large, erect fern, 3 to 6 feet tall.
- Compound leaf with incised pinnae arranged on each side of a leaf-axis like a feather.
- Fruiting bodies arranged in continuous, short, straight series resembling a chain.

BIBLIOGRAPHY

Dunne, T., L.B. Leopold. 1978. *Water in Environmental Planning*. W.H. Freeman, New York. 818 pp.

Howell, J.T. 1970. *Marin Flora*. University of California Press, Berkeley, CA. 366 pp.

Jones, R.R., D.A. Jones (tech. coords.). 1977. Importance, preservation, and management of riparian habitat: a symposium. (Tuscon, AZ, July 9, 1977) USDA Forest Service, Rocky Mountain Forest and Range Experiment Station, Fort Collins, CO. General Technical Report GTR-RM-43. 217 pp.

Mason, H.L. 1969. *A Flora of the Marshes*. University of California Press, Berkeley, CA. 878 pp.

McMinn, H.E., E. Maino. 1967. *An Illustrated Manuel of Pacific Coast Trees*. University of California Press, Berkeley, CA. 408 pp.

Munz, P. A. and D.D. Keck. 1973. *A California Flora*. University of California Press, Berkeley, CA. 1912 pp.

Warner, R.E., K.M. Hendrix. 1984. *California Riparian Systems, Ecology, Conservation, and Productive Management*. University of California Press, Berkeley, CA. 1035 pp.

Warner, R.E., K.M. Hendrix. 1985. *Riparian Resources of the Central Valley and California Desert*. California Department of Fish and Game, Sacramento, CA. 145 pp.

GLOSSARY

akene - dry, indehiscent, 1-seeded fruit from a 1-chambered ovary, often appearing as a naked seed as in a sunflower seed.

annual - a plant that lives for one year or growing season.

apetalous - without petals.

bract - a reduced or modified leaf, particularly the scale-like leaves in an inflorescence.

calyx - outermost or lowermost whorl of flowerparts, the sepals collectively.

catkin - a deciduous spike or spike-like inflorescence of unisexual apetalous flowers as in walnut, birch, or willow.

compound leaf - a leaf separated into two or more distinct leaflets.

corolla - whorl of flower parts immediately inside or above the calyx, the petals collectively; often large and brightly colored.

deciduous - losing leaves seasonally.

decumbent - reclining on the ground but with the growing end ascending.

dentate - pertaining to leaf margins having teeth that point outward.

drupe - fleshy or pulpy, indehiscent berry-like fruit in which a single seed is encased in a stone as in a cherry.

entire - pertaining to leaf margins that are continuous and smooth.

epiphytic - without soil roots; obtaining nutrients from air, rainwater and organic debris.

exfoliating - peeling off in thin layers as in an onion skin.

follicle - dry, usually many-seeded fruit, opening along only one side as in milkweed.

head - a compact group of flowers that occur in a cluster at the end of a stem or branch as in a sunflower.

herbaceous - pertaining to a plant without a persistent woody stem as in lettuce; a plant that dies back below groud each year.

indehiscent - not opening to release contents; usually pertaining to fruits like acorns.

indusium- in many ferns, a veil- or scale-like outgrowth of the leaf that covers a sorus.

inflorescence - the arrangement of flower clusters on a floral axis.

leaf axil - the upper angle between a leaf and a stem.

palmate - radiating from a common point as fingers radiate from one's palm; usually pertaining to veins, lobes, of leaflets of a leaf.

panicle - a compound inflorescence branching from a central axis.

perennial - a plant that lives for more than 2 years or growing seasons.

pinnate - a compound leaf with paired leaflets on opposite sides of an elongated axis as in a feather.

pome - a fleshy indehiscent fruit as in an apple.

raceme - an unbranched inflorescence with a single axis, the flowers arranged on simple stalks as in lupines.

rhizome - an horizontal, often elongate, underground stem as in asparagras.

riparian - streamside, an association of plant species growing adjacent to fresh-water courses, including perennial and intermittent streams, lakes and other bodies of fresh water.

sheath - a tubular basal part of a leaf that wraps around a stem.

spadix - a floral spike enveloped in a spathe.

spathe - a broad sheathing bract enclosing a spadix as in a calla.

spike - a type of inflorescence with numerous stemless flowers growing directly from an elongated axis.

spikelets - segments of a grass inflorescence surrounded by a pair of bracts.

stigma - the pollen-receiving part of a female flower.

stipules - a leaf- or scale-like appendage at the base of a leaf stem (petiole).

umbel - inflorescence of few to many flowetrs on stalks of approsimately equal length arising from the top of a flower stem.

FIGURE 1

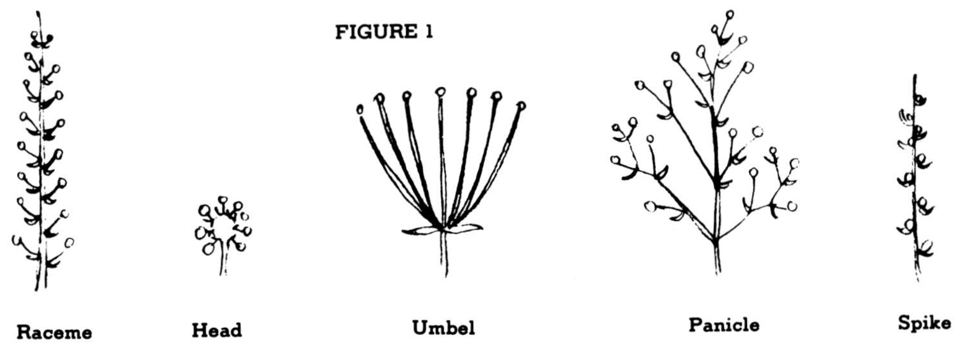

INDEX

ABOUT THE AUTHORS

Phyllis M. Faber lives close to San Francisco Bay in Mill Valley, California. As a biologist with degrees in Zoology and Microbiology, she has explored, taught, and worked in the coastal marshes and wetlands California for many years. The protection and preservation of these wetlands, and of California's natural and agricultural resources have been a major focus of her professional work and conservation activities. She is currently editor of *Fremontia*, the journal of the California Native Plant Society, a part-time faculty member of the Marin Community College, and has been monitoring a major salt marsh restoration project in San Francisco for the Golden Gate Bridge, Highway, and Transportation District. Her first book in this series of publications, *Common Wetland Plants of Coastal California*, won an outstanding environmental document award for 1982 from the Association of Environmental Professionals and a special mention from the prestigious Bookbuilders West for the innovative use of xerography. She is a co-author of the U.S. Fish and Wildlife publication *Riparian Habitats of the Southern California Coastal Region: A Community Profile.*

Robert F. Holland was introduced to rivers and their ways long before he began his formal training as a plant ecologist. He spent his boyhood playing along the lower American River in Sacramento County, California. During high school and college he had summer jobs as a river guide on many of the larger and smaller streams of the American west and dabbled, without distinction, in competitive white water kayaking. After undergraduate training in biology and earth science at Antioch College in Ohio, he returned to California, where he won his doctorate in ecology from the University of California at Davis in 1978. He is currently Vegetation Ecologist, under contract to The Nature Conservancy, working in the Non-game Heritage Program of the California Department of Fish and Game. Since 1978, he has studied vegetation from alpine summits to tropical lowlands, but has continually found his way back to the shade along some slow-moving Great Valley stream, where he finds great strength in just sitting and watching the river flow.

NOTES

NOTES

NOTES